Advances in 21st Century Human Settlements

Indexed by SCOPUS

This Series focuses on the entire spectrum of human settlements – from rural to urban, in different regions of the world, with questions such as: What factors cause and guide the process of change in human settlements from rural to urban in character, from hamlets and villages to towns, cities and megacities? Is this process different across time and space, how and why? Is there a future for rural life? Is it possible or not to have industrial development in rural settlements, and how? Why does 'urban shrinkage' occur? Are the rural areas urbanizing or is that urban areas are undergoing 'ruralisation' (in form of underserviced slums)? What are the challenges faced by 'mega urban regions', and how they can be/are being addressed? What drives economic dynamism in human settlements? Is the urban-based economic growth paradigm the only answer to the quest for sustainable development, or is there an urgent need to balance between economic growth on one hand and ecosystem restoration and conservation on the other – for the future sustainability of human habitats? How and what new technology is helping to achieve sustainable development in human settlements? What sort of changes in the current planning, management and governance of human settlements are needed to face the changing environment including the climate and increasing disaster risks? What is the uniqueness of the new 'socio-cultural spaces' that emerge in human settlements, and how they change over time? As rural settlements become urban, are the new 'urban spaces' resulting in the loss of rural life and 'socio-cultural spaces'? What is leading the preservation of rural 'socio-cultural spaces' within the urbanizing world, and how? What is the emerging nature of the rural-urban interface, and what factors influence it? What are the emerging perspectives that help understand the human-environment-culture complex through the study of human settlements and the related ecosystems, and how do they transform our understanding of cultural landscapes and 'waterscapes' in the 21st Century? What else is and/or likely to be new vis-à-vis human settlements – now and in the future? The Series, therefore, welcomes contributions with fresh cognitive perspectives to understand the new and emerging realities of the 21st Century human settlements. Such perspectives will include a multidisciplinary analysis, constituting of the demographic, spatio-economic, environmental, technological, and planning, management and governance lenses.

If you are interested in submitting a proposal for this series, please contact the Series Editor, or the Publishing Editor:

Bharat Dahiya (bharatdahiya@gmail.com) or
Loyola D'Silva (loyola.dsilva@springer.com)

Vishal Chettry · Harsimran Kaur ·
Sarbeswar Praharaj · Bharat Dahiya
Editors

Geoinformatics for Hilly Regions

Sustainable Human Settlements

Springer

Editors
Vishal Chettry
Department of Architecture, Planning and Design
Indian Institute of Technology (BHU)
Varanasi, Uttar Pradesh, India

Harsimran Kaur
Department of Architecture, Planning and Design
Indian Institute of Technology (BHU)
Varanasi, Uttar Pradesh, India

Sarbeswar Praharaj
Department of Geography and Sustainable Development and School of Architecture
University of Miami
Coral Gables, FL, USA

Bharat Dahiya
School of Global Studies, Research Center for Sustainable Development and Innovation
Thammasat University
Bangkok, Thailand

School of Public Leadership
Stellenbosch University
Stellenbosch, Western Cape, South Africa

ISSN 2198-2546 ISSN 2198-2554 (electronic)
Advances in 21st Century Human Settlements
ISBN 978-981-95-0155-7 ISBN 978-981-95-0156-4 (eBook)
https://doi.org/10.1007/978-981-95-0156-4

This Springer imprint is published by the registered company Springer Nature Singapore Pte Ltd.
The registered company address is: 152 Beach Road, #21-01/04 Gateway East, Singapore 189721, Singapore

This book is dedicated to all the researchers, practitioners, and policymakers who tirelessly work toward the sustainable development of hilly regions. Their commitment to the preservation of natural landscapes, the resilience of human settlements, and the integration of geoinformatics in addressing complex challenges inspires our collective efforts.

We also dedicate this work to our families and loved ones for their unwavering support and encouragement throughout this journey.

—Dr. Vishal Chettry, Dr. Harsimran Kaur, Dr. Sarbeswar Praharaj, and Prof. Bharat Dahiya

Foreword

It is my pleasure to write the foreword to this significant volume, Geoinformatics for Hilly Regions: Sustainable Human Settlements, co-edited by Dr. Vishal Chettry, Dr. Harsimran Kaur, Dr. Sarbeswar Praharaj, and Prof. Bharat Dahiya. This book addresses a critical and timely subject, i.e., the sustainable development of human settlements in hilly regions. This topic demands urgent attention from scholars, practitioners, and policymakers alike.

Hilly regions, with their unique environmental, cultural, and infrastructural characteristics, present distinct challenges that conventional urban planning and development strategies often fail to address adequately. The steep topography, fragile ecosystems, and climatic variations typical of these areas require innovative approaches to ensure that human settlements are both resilient and sustainable. This book, through its comprehensive and multidisciplinary approach, offers valuable insights into the specific needs of hilly regions and provides practical solutions that can be adapted to similar environments worldwide.

The editors have brought together a diverse group of experts, whose collective knowledge spans the fields of urban planning, environmental science, geoinformatics, and sustainable development. The result is a volume that not only explores the theoretical underpinnings of sustainable urban development in hilly regions but also presents empirical research and case studies that illustrate the practical application of these theories. This blend of theory and practice is one of the book's greatest strengths, offering readers both the conceptual frameworks needed to understand the complexities of hilly region development and the concrete examples needed to implement sustainable solutions.

One of the key themes that this book explores is the role of geoinformatics, a field that has become increasingly important in urban and environment planning. The use of remote sensing, geographic information systems (GIS), and other geospatial technologies allows for a more precise understanding of the challenges faced by hilly regions. From monitoring air quality to assessing the impacts of climate change, the applications of geoinformatics in this context are vast and varied. Another important aspect of this book is its emphasis on the cultural and social dimensions of sustainable development in hilly regions. The editors and contributors recognize that sustainable urban development is not just about managing physical resources, it is also about understanding and respecting the cultural and social contexts in which development occurs. This volume explores how cultural practices, community values, and social dynamics influence the way in which hilly regions are developed, and how these factors can be integrated into sustainable planning processes.

In addition to its scholarly contributions, the book "Geoinformatics for Hilly Regions: Sustainable Human Settlements" is also a call to action. As the world faces increasing environmental challenges, the need for sustainable solutions has never been greater. It is my hope that this volume will inspire further research and innovation in the field of sustainable urban development, particularly in regions that are often overlooked in mainstream urban studies.

The editors deserve special recognition for their dedication and perseverance in bringing this volume to fruition. Their ability to coordinate the efforts of contributors from different countries and disciplines is a testament to their commitment to the cause of sustainable development. I also extend my gratitude to the contributing authors, whose research and insights form the foundation of this book. I am confident that this book will serve as a reference for academics, professionals, and policymakers, guiding them in their efforts to create resilient, sustainable, and inclusive human settlements in some of the most challenging and beautiful environments on Earth.

Prof. Mark Roseland
School of Community Resources and Development (CRD)
4020 Watts College of Public Service and Community Solutions
Arizona State University Phoenix
Phoenix, AZ, USA

Other Affiliation:
Affiliate Faculty (thru AY 2024)
School of Geographical Sciences and Urban Planning
Watts College of Public Service and Community Solutions
Senior Global Futures Scientist
Global Futures Scientists and Scholars
e-mail: mark.roseland@asu.edu

Preface

This anthology, *Geoinformatics for Hilly Regions: Sustainable Human Settlements*, emerged from a shared vision for sustainable urban development in the hilly regions of India. This collaborative effort brings together editors based in different countries: Dr. Vishal Chettry and Dr. Harsimran Kaur in India, Dr. Sarbeswar Praharaj in the USA, and Prof. Bharat Dahiya who holds academic positions in Thailand and South Africa. Each editor contributed their unique perspectives and interdisciplinary expertise to address the distinctive challenges and opportunities observed for the sustainable development of human settlements in the hilly regions.

Our journey of collaboration has been remarkable, showcasing the power of modern technology and a collective dedication to sustainability. Despite geographical distances, we worked seamlessly through numerous virtual meetings to align our goals, structure the content, and delineate responsibilities. These sessions were pivotal in fostering a cohesive vision and ensuring a comprehensive approach to the subject matter. Our shared vision was developed over the summer of 2023 and resulted in the signing of book contract with Springer Nature in September 2023.

The process of preparing this co-edited book involved reaching out to authors who were (and are) passionate researchers in the field of sustainable planning and design for hilly settlements. These experts contributed their research, forming the empirical foundation of this volume. Their in-depth studies and practical insights have enriched the book, providing a thorough understanding of various aspects of sustainable urban development in hilly regions.

The significance of this topical anthology lies in its holistic approach to sustainable human settlements development in hilly regions. It covers a wide range of themes, including concepts of sustainable urban development, comparative studies on land use planning, and sustainable land-use practices in environmentally sensitive areas. The impact of culture, climate, topography, and geology on settlements is explored, along with the challenges and opportunities for accessibility for senior citizens in hill settlements. Further, the book delves into the dynamics of urban growth, utilizing remote sensing techniques for air quality monitoring and spatio-temporal analysis of environmental parameters during significant events, like the COVID-19 pandemic. It also addresses climate change vulnerability assessments, tourism vulnerability mapping, and sustainable disaster management strategies. All of this contributes toward normative priorities outlined in the "New Urban Agenda" [2], "the 2030 Agenda for Sustainable Development" [1], and the "Sustainable Development Goal (SDG) 11" [3].

We extend our heartfelt gratitude to the contributing authors, whose expertise and dedication were instrumental in shaping this collective project. Their systematic research and insightful manuscripts have significantly enriched the book's content. For their scholarly contributions, we extend our gratitude to Sonal Atreya, Sarika Bahadure, Swati Bahale, Nivedita Barman, Prabhjot Singh Chani, Vishal Chettry, Bharat Dahiya, Amrita Dwivedi, Sameer Gujar, Jit Kumar Gupta, Minakshi Jain, Tanushri Mokde Kamble, Harsimran Kaur, Ashwani Kumar, Vaibhav Kulkarni, Dilip H. Lataye, Keerti Manisha, Amit Maurya, Akhil Nawani, Apoorva Nawani, Sarbeswar Praharaj, Parshant Rehal, Sneha, Ashutosh Saini, Divyanshu Saini, Bhavna Shrivastava, Shalini Sharma, Shubham Sharma, Swasti Sharma, Inderpal Singh, Madan Lal Suthar, Ranjeet Verma, and Avinash Yadav for their significant contributions to this volume.

Special thanks are due to our chapter reviewers: Sayeed Alig, Arghadeep Bose, Vishal Chettry, Akshay Klasareswar, Harsimran Kaur, Meenal Kumar, Rituraj Neog, Prasanta Patra, and Sarbeswar Praharaj for their meticulous peer reviews and constructive feedback. Their expertise and critical insights have significantly enhanced the quality and rigor of the various chapters.

The Editors pay sincere regards to Prof. Mark Roseland from School of Community Resources and Development, Senior Global Futures Scientist, Global Futures Laboratory, Arizona State University, USA, for writing a thought-provoking foreword to this co-edited book.

We also wish to acknowledge Prof. Bharat Dahiya, co-editor of this volume as well as Series Editor of the Springer Nature book series, *Advances in 21st Century Human Settlements*, for his continuous guidance throughout the editing and publishing process. His insights and advice have been invaluable, ensuring that the book meets the highest standards of academic rigor and relevance.

Taking this opportunity, we sincerely acknowledge, with sincere thanks, the institutional support provided by our respective institutions: Department of Architecture, Planning and Design (APD), Institute of Indian Technology (Banaras Hindu University), Varanasi, India; Department of Geography and Sustainable Development and School of Architecture, University of Miami, USA; and School of Global Studies, Thammasat University, Bangkok, Thailand, and School of Public Leadership, Stellenbosch University, South Africa, for providing us with all necessary support and facilities from the inception of this co-edited book to its publication.

Lastly, we are also deeply appreciative of our publishing team from Springer Nature, especially Dr. Loyola D' Silva (Executive Editor), Ms. Shalini Monica Clement Selvam (Project Coordinator—Total Service), and Ms. Niyati Naudiyal (Editorial Assistant, Science and Technology), for their unwavering support and professionalism, ensuring the seamless production of this volume.

This book is not just a scholarly endeavor but a labor of love and commitment to sustainable urban development in hilly regions. We hope it captivates academic communities, professionals, and the general public alike, serving as a valuable resource that inspires future research and practical action in this critical area of sustainable human settlements development in the twenty-first century.

Varanasi, India — Dr. Vishal Chettry
Varanasi, India — Dr. Harsimran Kaur
Miami, USA — Dr. Sarbeswar Praharaj
Bangkok, Thailand and Stellenbosch, South Africa — Prof. Bharat Dahiya

References

1. United Nations (2015) Transforming our World: The 2030 Agenda for sustainable development, A/RES/70/1. https://sustainabledevelopment.un.org/post2015/transformingourworld/publication
2. United Nations (2017) The New Urban Agenda, A/RES/71/256, Habitat III and United Nations. https://habitat3.org/the-new-urban-agenda/
3. United Nations (2024) Sustainable development Goal 11: make cities and human settlements inclusive, safe, resilient and sustainable. https://sdgs.un.org/goals/goal11

Contents

Editors and Contributors

About the Editors

Vishal Chettry is currently working as an Assistant Professor at the Department of Architecture, Planning and Design, IIT (BHU) Varanasi (an Institute of National Importance, Government of India). He was awarded a Ph.D. in "Delineation of Urban Growth Boundary to Mitigate Urban Sprawl in Indian Mid-sized Cities" from VNIT Nagpur (an Institute of National Importance, Government of India) in 2021. Before his current role, Dr. Chettry served as an Assistant Professor at Manipal School of Architecture and Planning, MAHE Manipal (an Institute of Eminence declared by Government of India) from July 2021 to March 2023. He holds a Master of Planning (Urban and Regional Planning) from the School of Planning and Architecture Bhopal (an Institute of National Importance, Government of India) in 2016 and a Bachelor of Architecture in 2014 from Rashtrasant Tukadoji Maharaj Nagpur University (RTMNU) formerly Nagpur University.

Dr. Chettry's accomplishments include receiving a certificate of appreciation for the best thesis project titled "City Center at Guwahati" from the Indian Institute of Architects (IIA) Nagpur chapter in 2014. He has published research papers in peer-reviewed international journals of repute and presented research findings at various national and international conferences. Furthermore, he has successfully organized international conferences, workshops, and multiple training programs to promote sustainable urban development. He has delivered lectures on Low-Cost Materials for Construction, Vernacular Architecture, and Strategies for Architectural Thesis Success at various forms of national repute. He has been invited as a peer reviewer by many international journals and conferences.

He has been Principal Investigator of research projects funded by Indian Council for Social Science Research (ICSSR) New Delhi, and IIT(BHU), Varanasi. He has also collaborated with National Institute of Disaster Management (NIDM), New Delhi, to organize a 03-day National Training Program on Water Urbanism for Resilient Development: Case Studies from Varanasi, India. He has completed consultancy projects funded by Northern Railways, Government of India and National

Monument Authority, Ministry of Culture, Government of India. He is also an editorial board member of the journal "Scientific Reports" and "Discover Sustainability" published by Springer Nature. His research interests include urban sprawl, urban sustainability, urban heat island, and the application of RS and GIS in the urban and regional planning domain.

Dr. Chettry can be contacted via email at vishalchettry33@gmail.com or vishal.apd@iitbhu.ac.in

Harsimran Kaur is a certified Indian Green Building Council Accredited Professional (IGBC-AP) and Assistant Professor at the Department of Architecture, Planning and Design at IIT (BHU), Varanasi. She specializes in Urban Sustainability, Hill Settlements, Architectural Heritage Conservation, Virtual and Augmented Reality, Spatial Analytics, and Data Visualization. Dr. Kaur completed her Ph.D. in June 2020 from the Indian Institute of Technology Roorkee, where she also earned her M. Architecture in June 2015 with first division and distinction. She obtained her B. Arch in June 2011 from Punjab Technical University, achieving first division.

Dr. Kaur has held several notable positions, including her current role as Head of the Department since January 2024. She has been serving as an Assistant Professor at IIT (BHU) since July 2020. Previously, she was a Teaching Assistant at IIT Roorkee from January 2016 to December 2019 and an Assistant Professor at Lovely Professional University from July 2015 to December 2015.

Dr. Kaur has been involved in numerous consultancy and research projects, such as the creation of the IIT (BHU) Virtual Reality 360° Campus Tour funded by the IIT(BHU) Seed Grant and a study on the socio-economic impact of PM Awas Yojana(U) in Varanasi District sponsored by ICSSR. Other significant projects include heritage consultations for the National Monument Authority, a 3D mapping project using drones and GIS software funded by SERB-DST, and various heritage impact assessments and architectural consultancy works for organizations like Northern Railways, Ashirbad Constructions Pvt. Ltd., PSP Project Limited, and DCRUST Murthal.

Her extensive publication record includes articles in high-impact journals such as *Air Quality, Atmosphere and Health*, *Visions of Sustainability*, *Progress in Disaster Science*, *Environment, Development and Sustainability*, and *Journal of Urban Planning and Development*. Her work often focuses on urban sustainability, heritage conservation, and environmental analysis.

Dr. Kaur's achievements have been recognized with several awards, including being shortlisted for the IIA National Awards for Excellence in Architecture 2021, receiving the Outstanding Contribution by Young Professional Award from the Indian Concrete Institute in 2022, and being declared Champion at the national IITB-ISRO-AICTE Mapathon in 2021. She is also a certified Indian Green Building Council Accredited Professional and has received an International Travel Grant from IIT Roorkee for attending the SEB-2019 conference in Budapest, Hungary. Additionally, she was awarded the Senior Research Fellowship by the Ministry of Human

Resource Development for her Ph.D. at IIT Roorkee and was invited to the International Vancouver Summer Program at the University of British Columbia during her master's degree.

Dr. Kaur has organized and led several workshops and expert talks, such as the one-day workshop on the socioeconomic impact of PMAY(U) Project in Varanasi, a high-end workshop on 3D Mapping and Modeling using Drone and GIS Software, and a national training program on Water Urbanism for Resilient Development. She has also contributed to discussions on climate risk preparedness using intangible cultural heritage, sustainable hillside development, and the management of historic towns in South Asia.

Dr. Harsimran Kaur's contributions to the field of architecture and urban planning are noteworthy, and her research continues to impact urban sustainability and heritage conservation practices. She can be reached via email at harsimran.apd@iitbhu.ac.in

Dr. Sarbeswar Praharaj is the Director of Graduate Studies in Urban Sustainability and Resilience at the University of Miami, USA. He holds a tenure-track Assistant Professor appointment jointly with the Department of Geography and Sustainable Development and the School of Architecture. Dr. Praharaj is also an affiliate faculty with the University of Miami's flagship Climate Resilience Institute, where he combines interdisciplinary research and education at the intersection of urban science and climate resilience. Sarbeswar's research explores creative design and applications of urban geospatial decision-support technologies and platforms to support the advancement of community resilience planning and smart cities development. He teaches upper-level courses on urban analytics and climate-informed land use planning. Dr. Praharaj's recent research project titled "Simulating Coastal Adaptation and Local Exposure for Enhanced Resilience" has received funding from the US National Science Foundation.

Before joining UMiami, Sarbeswar led a 15-million-dollar university-wide research institute, "Knowledge Exchange for Resilience," at Arizona State University as its Associate Director. During his tenure at ASU, he was an Assistant Research Professor within the School of Geographical Science and Urban Planning. Dr. Praharaj received ASU's prestigious "Sun Award" for "Exemplary Service" and was elevated to a Senior Global Futures Scientist at the Julie Ann Wrigley Global Futures Laboratory. He worked closely with the internationally diverse PLUS Alliance Fellows across ASU, Kings College London, and UNSW Sydney, for convergent research design on smart, healthy, and accessible cities. His work at ASU has produced numerous urban resilience toolkits (e.g., Economic Resilience Dashboard, Household Expenditure and Income Gap Highlighter Tool (HEIGHT), and City of Phoenix Green Infrastructure Management Platform) and attracted significant competitive grant funding.

Sarbeswar obtained his Ph.D. from the Faculty of Built Environment, UNSW Sydney. He subsequently pursued Postdoctoral work with the UNSW City Futures Research Center, which is Australia's foremost center for urban studies. Dr. Praharaj was the inaugural Manager of the City Analytics Lab at UNSW, where he designed and experimented with cutting-edge urban technologies and data-driven analytics.

Sarbeswar was the Research and Development Manager for a machine-learning-based land valuation and infrastructure scenario modeling tool—RAISE Rapid Analytics and Interactive Scenario Explorer, which was later commercialized through a majority stake acquired by PEXA. Sarbeswar authored the Strathfield Smart City Roadmap for the greater Sydney city council. His work at UNSW received funding from the Australian Federal Government's Australia-India Council and Australia-Kora Foundation. He was bestowed with the UNSW International Postgraduate Research Scholarship for his Ph.D.

Dr. Praharaj can be reached via email at spraharaj@miami.edu

Bharat Dahiya is Director of the Research Center for Sustainable Development and Innovation at the School of Global Studies, Thammasat University, Bangkok, Thailand. He is an Extraordinary Professor at the School of Public Leadership, Stellenbosch University, Western Cape, South Africa. He is also a Founding Board Member of the World Smart Cities Economic Development Commission—under the World Business Angels Investment Forum, USA.

An award-winning urbanist, Bharat combines cutting-edge research, policy analysis, and development practice aimed at examining and tackling socio-economic, environmental and governance issues in the global urban context of sustainable development. Since the early-1990s, his research and professional work has focused on sustainable cities and urbanization, strategic urban planning and development, urban infrastructure, urban environment, climate change and urban resilience, post-disaster recovery and reconstruction, and cultural heritage and landscapes.

Working with the World Bank, UN-HABITAT, Asian Development Bank, UNDP, United Nations University, and Metropolis, Bharat has initiated, led and/or contributed to international development projects and programmes on sustainable urban development around the world. He has led the conceptualization, design and implementation a number of international development including the preparation of the first-ever "Citywide Pro-poor Ger-area Upgrading Strategy and Investment Plan for Ulaanbaatar", and "Community-Led Ger Area Upgrading in Ulaanbaatar City" (Mongolia); "Earthquake-resistant and Energy-efficient School Reconstruction in Sichuan Province", and "Getting Children Back to School: Providing Prefabricated Classroom Units to the Earthquake-Affected Children of Gansu Province" (China); and "City Development Strategy for Thanh Hoa City in a Regional Development Context" (Viet Nam). During 2012–2013, he led the preparation of "Ulaanbaatar Urban Services and Ger Areas Development Investment Program", a 10-year Multi-Tranche Financing Facility with a total lending of US $160 million, financed by the Municipal Government of Ulaanbaatar, the Asian Development Bank, and European Investment Bank.

Throughout his career, Bharat has been involved in the preparation and dissemination of cutting-edge knowledge products. At the World Bank headquarters in Washington DC, USA, he conducted the first-ever systematic review of the Bank's investments for improving urban liveability, published as a co-authored book, *Urban Environment and Infrastructure: Toward Livable Cities* (2004). At UN-HABITAT, he led, conceptualized, and coordinated the preparation of United Nations' inaugural report on *The State of Asian Cities 2010/11* (2010). He has also co-authored *Partnering for Sustainable Development: Guidelines for Multi–stakeholder Partnerships to Implement the 2030 Agenda in Asia and the Pacific* (UNU-IAS, and UN-ESCAP, 2018) and *Catalyzing Climate Finance with the Shandong Green Development Fund* (Asian Development Bank, 2020). He was the lead author of Metropolis's first-ever *Asian Metropolitan Report* (2021). He has co-edited six academic anthologies, *New Urban Agenda for Asia-Pacific: Governance for Sustainable and Inclusive Cities* (Springer Nature, 2020), *Practising Cultural Geographies: Essays in Honour of Rana PB Singh* (Springer Nature, 2022), *Disaster Resilience and Human Settlements: Emerging Perspectives in the Anthropocene* (Springer Nature, 2023), *City Responses to Disruptions in 2020: From Lockdowns to Aftermath* (Springer Nature, 2023), *Urban Health: A Global Perspective* (Elsevier, 2024), and *Urban Planning and Design for Megacities in the Global South: Smart and Sustainable Development* (Springer Nature, 2025).

Bharat advises several professional, academic, private sectors, and non-profit organizations around the world. Since 2019, he has been a Member of the International Advisory Board for the UN-HABITAT's *World Cities Report* (Nairobi, Kenya). Since early-2014, he has been the Series Editor for the SCOPUS-indexed Springer book series, *Advances in 21st Century Human Settlements*, which has a growing collection of over 50 monographs and edited volumes. He serves as a member of the editorial board of several international journals as well as the International Scientific Board of the book series, "Geographies of the Anthropocene", published by Il Sileno Edizioni, Italy. He is also a Member of the Advisory Group for "Future Earth Urban Knowledge–Action Network". He has held academic positions in Australia, Indonesia, South Africa, and Thailand.

In 1997, Bharat was elected a Fellow of the Cambridge Philosophical Society—the oldest scientific society in Cambridge, UK. In the same year, the Cambridge Commonwealth Trust admitted him to the status of Fellow of the Cambridge Commonwealth Society. At Cambridge, he was a Member of Sidney Sussex College, and a Ph.D. Research Scholar at the Department of Geography, under the Faculty of Earth Sciences and Geography.

For his professional contributions to sustainable urban development in Mongolia, the Government of Mongolia awarded Bharat with a "Certificate of Honour" (2009), and the Municipal Government of Ulaanbaatar decorated him with a "Medal of Honour" (2010). The Global Council for the Promotion of International Trade conferred on him a "Global Sustainability Award 2021".

Bharat holds a Ph.D. in "Urban Governance, Planning and Environment" from the University of Cambridge, United Kingdom. He completed his B.A. in Geography and Political Science from Maharshi Dayanand University, Haryana, India. He read for his M.A. in Geography from Jawaharlal Nehru University, and Master of Planning from School of Planning and Architecture, both based in New Delhi.

Contributors

Sonal Atreya Indian Institute of Technology, Roorkee, India

Sarika Bahadure Department of Architecture and Planning, Visvesvaraya National Institute of Technology, Nagpur, Maharashtra, India

Swati Bahale Department of Architecture, College of Engineering, Sungkyunkwan University, Suwon, Republic of Korea

Nivedita Barman Department of Environmental Science, Tezpur University, Tezpur, India

Prabhjot Singh Chani Indian Institute of Technology, Roorkee, India

Vishal Chettry Department of Architecture, Planning and Design, IIT (BHU) Varanasi, Varanasi, India

Bharat Dahiya School of Global Studies, Research Center for Sustainable Development and Innovation, Thammasat University, Bangkok, Thailand; School of Public Leadership, Stellenbosch University, Stellenbosch, Western Cape, South Africa

Amrita Dwivedi Department of Humanistic Studies, Indian Institute of Technology (Banaras Hindu University), Uttar Pradesh, Varanasi, India

Sameer Gujar School of Architecture and Planning, NICMAR University, Pune, Maharashtras, India

Jit Kumar Gupta IET Bhaddal, Chandigarh, India

Minakshi Jain Department of Architecture, National Institute of Technology Hamirpur, Hamirpur, Himachal Pradesh, India

Tanushri Mokde Kamble Institute of Architecture and Planning, Nirma University, Ahmedabad, Gujarat, India

Harsimran Kaur Department of Architecture, Planning and Design, Indian Institute of Technology (Banaras Hindu University), Uttar Pradesh, Varanasi, India

Vaibhav Kulkarni Padmabhushan Dr. Vasantdada Patil College of Architecture, Pune, Maharashtra, India

Ashwani Kumar Department of Architecture, National Institute of Technology, Hamirpur, India

Dilip H. Lataye Visvesvaraya National Institute of Technology, Nagpur, Maharashtra, India

Keerti Manisha Department of Architecture, National Institute of Technology Tiruchirappalli, Tiruchirappalli, Tamil Nadu, India

Amit Maurya Department of Architecture, Planning and Design, IIT(BHU), Varanasi, Uttar Pradesh, India

Akhil Nawani Department of Architecture, Planning and Design, Indian Institute of Technology (BHU) Varanasi, Varanasi, Uttar Pradesh, India

Sarbeswar Praharaj Department of Geography and Sustainable Development and School of Architecture, University of Miami, Coral Gables, FL, USA

Parshant Rehal Indian Institute of Technology, Roorkee, India

Ashutosh Saini School of Architecture and Design, Manipal University, Jaipur, India

Divyanshu Saini Visvesvaraya National Institute of Technology, Nagpur, Maharashtra, India

Shalini Sharma Indian Institute of Technology, Kharagpur, India

Shubham Sharma Indian Institute of Technology, Kharagpur, India

Swasti Sharma Amity School of Architecture and Planning, Amity University, Punjab, India

Bhavna Shrivastava Department of Architecture and Planning, Malaviya National Institute of Technology Jaipur, Jaipur, India

Inderpal Singh Department of Architecture, National Institute of Technology Hamirpur, Hamirpur, Himachal Pradesh, India

Sneha Department of Architecture, Planning and Design, IIT(BHU) Varanasi, Varanasi, India

Madan Lal Suthar Town Planning, Local Self Government, Government of Rajasthan, Jaipur, India

Ranjeet Verma Department of Humanistic Studies, Indian Institute of Technology (Banaras Hindu University), Uttar Pradesh, Varanasi, India

Avinash Yadav Department of Architecture, Planning and Design, IIT(BHU) Varanasi, Varanasi, India

Introduction: Geoinformatics for Sustainable Human Settlements in Hilly Regions

Vishal Chettry, Harsimran Kaur, Sarbeswar Praharaj, and Bharat Dahiya

1 Background

Human settlements in hilly regions signify a multifarious combination of environmental, social, and economic dynamics that offer challenges and opportunities for sustainable development. Unlike flat terrains, hilly landscapes pose unique obstacles due to their rugged topography, diverse microclimates, and susceptibility to natural hazards [1]. The mountain ecosystems have a significant impact on the sustainable development of the Himalayan region in India, as they possess half of the world's population with valuable goods and services and a diverse environmental heritage [2–5]. However, the impacts of unsustainable development are escalating and increasingly difficult to rectify within this fragile ecological context [6]. Given its geographical, social-cultural, economic, and ecological distinctiveness, research and development endeavours in this globally important mountainous system necessitate innovative approaches to urban planning, infrastructure development, and resource management to ensure the region's long-term sustainability.

In recent decades, the fundamental concepts, dimensions, and issues of sustainability have been rationally investigated and comprehended. However, the assessment of sustainability in environmentally sensitive hill regions and the consideration

V. Chettry (✉) · H. Kaur
Department of Architecture, Planning and Design, IIT (BHU) Varanasi, Varanasi, India
e-mail: vishal.apd@iitbhu.ac.in

S. Praharaj
Department of Geography and Sustainable Development and School of Architecture, University of Miami, Coral Gables, FL, USA

B. Dahiya
School of Global Studies, Research Center for Sustainable Development and Innovation, Thammasat University, Bangkok, Thailand

School of Public Leadership, University of Stellenbosch, Stellenbosch 7602, Western Cape, South Africa

V. Chettry et al. (eds.), *Geoinformatics for Hilly Regions*, Advances in 21st Century Human Settlements, https://doi.org/10.1007/978-981-95-0156-4_1

of site-specific and spatial natural environmental factors have been the focus of a limited number of academics, researchers, and institutions [7]. Furthermore, the consequences and impact of sustainability dimensions differ across various levels: micro-level, meso-level, and macro-level [8, 9]. Further, in order to assess the impact of urban and regional development on sustainability in hilly regions, the application of descriptive and deductive analytical methods and techniques is required, along with the integration of geoinformatics.

Geoinformatics—comprising Geographic Information Systems (GIS), remote sensing, and spatial analysis—offers a transformative solution to address the complexities of hilly terrain. Geoinformatics enables the collection, analysis, and visualization of spatial data, providing valuable insights into the interactions between human activities and the environment. By harnessing the power of geospatial technologies, stakeholders can make informed decisions to promote sustainable development in hilly regions.

2 Importance of Sustainable Human Settlements in Hilly Regions

Hilly regions around the world represent vital ecological and cultural landscapes, hosting a significant portion of the human population. These areas are characterized by rich and diverse ecosystems, ranging from lush forests to high-altitude grasslands, which support plenteous biodiversity and provide essential ecosystem services. Furthermore, hilly regions often hold cultural significance, serving as the ancestral homelands of indigenous communities and repositories of traditional knowledge and practices.

However, the sustainability of human settlements in hilly regions is increasingly threatened by rapid urbanization, population growth, and unsustainable land use practices. Unplanned development exacerbates environmental degradation, leading to deforestation, soil erosion, disturbance to natural drainage, pollution of streams and lakes, deforestation, desertification, and habitat loss. India has witnessed devastating floods in hilly regions, such as the 2013 Kedarnath floods in Uttarakhand, where a cloudburst and glacial outburst destroyed settlements and claimed thousands of lives [10]. Similarly, the 2018 Kerala floods, exacerbated by deforestation and unplanned development, submerged hilly areas, causing massive loss of life and property [11]. Further, in 2024, landslides caused by heavy rain killed more than 100 people in the Wayanad district of Kerala [12].

The construction of modern buildings not only neglects the importance of harmonizing such structures with the traditional and indigenous built environment and architectural integrity but also significantly contributes to environmental degradation by employing non-local materials, unconventional building methods, and design criteria that deviate from conventional standards. This incorporates a diverse array of endeavours, including the deforestation process for wood and the clearance of hill slopes. The prevalence of microclimatic changes, surface and subsurface hydrology disruptions, landslides, and soil erosion has increased, consequently leading to a rise in the occurrence of pollution, siltation, and building collapses causing loss of lives, livelihoods, and cultural and other properties. Water insecurity is caused by changes such as climate-induced changes such as diminished precipitation, depletion of natural springs, reduction in forest cover, and increased temperatures. The overall environmental quality is severely undermined by the heightened construction activity in numerous hill communities, which exacerbates these environmental issues [13].

The imperative to achieve sustainable human settlements in hilly regions has never been more pressing. Sustainable development involves aligning economic growth with environmental protection and social equity, ensuring that current needs are fulfilled without compromising the ability of future generations to meet their own needs. By adopting holistic approaches to urban planning and natural resource management, stakeholders can mitigate the adverse impacts of unplanned, haphazard and unsustainable urbanization, and safeguard the ecological integrity and cultural heritage of hilly terrains [14, 15].

Geoinformatics provides a comprehensive framework for understanding, analysing, and managing the complex interactions between humans and their natural environment in hilly regions. By leveraging large quantities of spatial data, advanced analytical techniques, and visualization tools, geoinformatics enables stakeholders to make informed decisions towards sustainable development. From land use planning to disaster risk reduction, geoinformatics offers a holistic approach to address the challenges faced by human settlements in hilly regions.

3 Applications of Geoinformatics in Hilly Regions

3.1 About Geoinformatics

Geoinformatics, also known as geospatial informatics or geographic information science (GIScience), is a dynamic interdisciplinary field that integrates geospatial technologies with information science [16]. Geoinformatics has emerged as a significant tool for decision-makers across various disciplines, industries, and organizations [17, 18]. Geoinformatics combines 'geo'—pertaining to the Earth, with 'informatics'—the study of information processing. Essentially, geoinformatics merges Earth sciences with Information Technology (IT). It encompasses the utilization of IT for gathering, analysing, storing, retrieving, representing, and disseminating information pertaining to the Earth.

3.2 *Components of Geoinformatics*

The components of geoinformatics typically include Geographic Information Systems (GIS), Remote Sensing, Global Positioning System (GPS), Cartography, Photogrammetry, Spatial Analysis, Database Management, and Geo-visualization [14]. GIS is the core component of geoinformatics. It is a system designed to capture, store, manipulate, analyse, manage, and present spatial or geographic data through software. Remote sensing involves the collection of data about the Earth's surface from a distance, typically using satellites, aircraft, or drones equipped with sensors. Remote sensing data includes imagery (such as aerial photographs and satellite images) and non-image data (such as temperature, precipitation, and vegetation indices). GPS is a satellite-based navigation system that provides location and time information anywhere on Earth where there is an unobstructed line of sight to four or more GPS satellites. GPS receivers can accurately determine the latitude, longitude, and elevation of a specific point on the Earth's surface. GPS data is commonly used for mapping, navigation, asset tracking, and georeferencing spatial data. Cartography is the art and science of mapmaking. It involves the design, creation, and interpretation of maps to communicate spatial information effectively. Cartographic principles are applied to create maps that are visually appealing, informative, and easy to understand. Photogrammetry is a technological advancement geared towards extracting geometric characteristics of objects by analysing their photographic representations. Its primary focus lies in gauging the positioning and structural attributes of objects through the utilization of photographs. Spatial analysis involves the examination, interpretation, and manipulation of spatial data to extract meaningful patterns, relationships, and insights. It includes overlay analysis, proximity analysis, interpolation, surface modelling, network analysis, and spatial statistics. Database Management Systems (DBMS) is used to store and manage spatial data efficiently. It is also employed to organize, index, query, and update geographic datasets. Spatial databases support operations such as spatial querying, spatial indexing, and spatial data processing. Geo-visualization refers to the graphical representation and visualization of geographic data. It involves the creation of maps, charts, graphs, and interactive visualizations to explore and communicate spatial patterns, trends, and relationships. Geo-visualization techniques help users to interpret complex spatial data and gain insights into geographic phenomena. All of these components work together to support various applications in fields such as urban planning, environmental management, natural resource assessment, transportation planning, disaster management, agriculture, and public health [19].

3.3 Application of Geoinformatics to Human Settlements and Sustainable Development

In recent years, the application of geoinformatics has revolutionized the understanding and management of diverse geographical landscapes, with particular significance in hilly regions [20]. Characterized by rugged terrain and unique environmental challenges, these areas demand tailored approaches for effective study and sustainable development. Geoinformatics, encompassing tools such as GIS and remote sensing technologies, offers a comprehensive toolkit to address the complexities of hilly landscapes. From terrain analysis to natural resource management and disaster risk assessment, the integration of geoinformatics techniques enables researchers and policymakers to gain valuable insights and make informed decisions [21–23]. Terrain analysis includes assessing slope gradients, identifying aspects, and profiling elevation, all of which are essential for understanding the intricate landforms characteristic of hilly terrain [24]. Such analyses contribute to evaluating factors like erosion susceptibility and the potential for landslides, crucial for effective land management and hazard mitigation strategies [25, 26]. Another critical application is in the domain of natural resource management [27]. Geoinformatics enables the assessment and monitoring of vital resources such as water, forests, and minerals in hilly areas. Through the integration of remote sensing data and GIS techniques, researchers can map resource distribution, monitor changes over time, and develop strategies for sustainable resource utilization. Additionally, geoinformatics supports habitat mapping and biodiversity conservation initiatives, contributing to the preservation of unique ecosystems found in hilly regions [28].

Furthermore, geoinformatics serves as a crucial tool for environmental monitoring and conservation efforts in hilly regions which includes assessment of land cover changes, vegetation dynamics, and habitat fragmentation over time [29]. This information is invaluable for identifying areas of ecological importance and prioritizing conservation initiatives aimed at preserving biodiversity and ecosystem services. It also facilitates the integration of spatial data from multiple sources, including satellite imagery, aerial surveys, and ground-based monitoring stations, enabling comprehensive assessments of environmental health and resilience in hilly landscapes [30]. Through the analysis of spatial patterns and trends, geoinformatics supports evidence-based decision-making for sustainable land management and conservation planning, ensuring the long-term viability of these ecologically sensitive areas.

Moreover, geoinformatics has played a vital role in supporting community engagement and participatory decision-making processes in hilly regions. By providing tools for spatial data visualization and analysis, GIS technology enables stakeholders to better understand the spatial implications of proposed development projects, land use changes, and natural resource management strategies [31]. Several communities have been empowered with spatial information and analytical capabilities, geoinformatics facilitates informed decision-making processes that reflect the diverse interests and perspectives present in hilly landscapes, ultimately promoting more equitable and sustainable development outcomes.

Finally, geoinformatics plays a crucial role in disaster risk management [32]. By leveraging spatial data and modelling techniques, researchers have assessed the vulnerability of hilly areas to natural hazards such as landslides, floods, and earthquakes. This information is invaluable for developing early warning systems, emergency response plans, and land-use policies aimed at reducing risks and enhancing resilience in these dynamic landscapes. Overall, the application of geoinformatics in studies focusing on hilly regions is instrumental in understanding their complexities and supporting sustainable development efforts.

4 Overview of the Book

This book titled 'Geoinformatics for Hilly Regions: Sustainable Human Settlement' aims to bridge critical gaps in scholarly exploration pertaining to investigation of human settlements in hilly regions of India to promote sustainable development. With a comprehensive exploration of various domains including land use planning, environmental sensitivity, mobility challenges, urban growth dynamics, air quality monitoring, and climate change vulnerability assessment, this book offers valuable insights into the unique challenges faced by hilly regions.

From the comparative study of land use planning in cities and human settlements from across the hilly regions of India to the assessment of urban growth dynamics in specific regions like Srinagar, Jammu and Kashmir, the book highlights various approaches for sustainable human settlements development in hilly regions. Moreover, it emphasizes the significance of incorporating cultural, climatic, topographical, and geological factors into settlement planning, as demonstrated in case studies like Gangtok and Kandad. Furthermore, the book explores the role of modern technologies such as satellite remote sensing and geoinformatics in monitoring air quality, assessing climate change vulnerability, and mapping tourism vulnerability. It also addresses critical issues like sustainable disaster management and inclusive habitat development in hilly regions, particularly focusing on the capital city and district of Shimla district in the state of Himachal Pradesh. Through its diverse array of chapters, the book underscores the necessity of holistic and context-specific approaches to sustainable urban development in hilly terrains, offering valuable insights, options, and strategies for creating inclusive, safe, resilient, and sustainable habitats in these regions. This edited volume includes fourteen chapters wherein case studies from the various hilly regions in India.

The first chapter titled 'Introduction: Geoinformatics for Sustainable Human Settlements in Hilly Regions', co-authored by the Chettry et al. [33] introduces the theme of the book. It presents an overview to the focus of this book which includes human settlements, hilly regions, application of geoinformatics in hilly regions.

The second chapter titled 'Sustainable Urban Development in Hilly Regions: Concepts, Evolution, and Challenges', co-authored by Kaur and Nawani [34] explores the use of GIS and remote sensing technologies to tackle the unique challenges of sustainable urban development in hilly regions. The diverse possibilities of remote sensing, highlighting its ability to collect spatial data on topographical features, vegetation health, and land types, are examined. Further, it focuses on identifying high-risk locations through remote sensing data to recommend preventive measures. Then GIS-based land use planning is investigated, highlighting the significance of maintaining ecological integrity. Additionally, the research is conducted into the use of remote sensing in watershed management to improve water resource management and prevent floods. The chapter provides an analysis of GIS-enabled infrastructure planning for resilient and sustainable development. It does so by incorporating case studies and examples demonstrating successful implementations of these technologies in sustainable urban development throughout the chapter. Challenges and factors to be taken into account when implementing GIS and remote sensing are examined, along with suggestions for future advancements and interdisciplinary collaborations.

The third chapter titled 'Land Use Planning in Hill Cities of India and Abroad: A Comparative Study', co-authored by Sharma et al. [35] provides an overview of land use planning in hill areas of India to better understand the existing development conditions in the hill towns. The land use practices followed in the hill towns of India are studied with reference to three case studies namely, Bilaspur in Himachal Pradesh, New Tehri in Uttarakhand, and Lavasa in Maharashtra. To evaluate the corresponding situation of land use planning practices in hill areas at international level, the case studies of three countries namely, Bhutan, Nepal and the United States are conducted. In addition, a comparative study of best land use practices being followed in hill towns in the country and abroad is also carried out, for identifying any lacunas in land use planning techniques adopted for hill towns in India. Further, an attempt is made to explicitly highlight difference between the process followed in the similar fragile areas.

The fourth chapter titled 'Sustainable land-use planning in the environmental sensitive hill city of Gangtok, India' co-authored by Bahadure et al. [36] analyses the impact of the unique geographical location, topographical features, and abundance of natural resources on hill cities. The unchecked pace of urbanization coupled with inadequate planning regulations has resulted in urban sprawl, disruption of natural landscapes, and exploitation of natural resources, ultimately steering hill cities towards an unsustainable future. In India, urban planning primarily revolves around megacities, leaving hill cities overlooked and underserved. Addressing the distinctive issues faced by hill cities necessitates a strategic approach to spatial planning. This chapter unveils a methodology tailored for conducting land suitability analysis, specifically focusing on Gangtok, India. The study aims to craft a spatial planning blueprint conducive to sustainable development, meticulously considering the physical, social, and environmental facets unique to Gangtok. The outlined methodology furnishes a robust framework adaptable for land use planning and fostering sustainable futures, while accounting for the carrying capacity of hill cities.

The fifth chapter titled 'The impact of culture, climate, topography, and geology on settlements in hilly areas: A case of Kandad, India', co-authored by Gujar et al. [37] examines the sustainability of settlement in Kandad Village, situated in the mountainous region of the Shivalik range, Chakrata town, in Uttarakhand state in

India, through the lens of GIS analysis. With a population of 265 residents and covering an area of three square kilometres, Kandad Village presents an ideal setting to explore sustainable development challenges within mountainous regions. Utilizing GIS techniques, the study conducts geographical and topological analyses to map the physical landscape and identify vulnerable areas prone to natural hazards. Furthermore, built-up change analysis is employed to understand land-use dynamics over time. The chapter also emphasizes the importance of traditional construction techniques in achieving resilient development by documenting and analysing structures within the village. Integration of GIS findings with traditional construction documentation informs discussions on sustainable settlement planning and development. The chapter concludes with reflections on the implications of the study's findings and outlines limitations and future research directions.

The sixth chapter titled 'Accessibility for Senior Citizens in Hill Settlements: Challenges and Opportunities for Sustainable Mobility', co-authored by Rehal et al. [38] analyses the challenges that older adults residing in urban hill settlements face towards their outdoor mobility and accessibility experiences. The chapter underscores the need for this research study by focussing on the projected population ageing future for most developing nations, including India, and changing family structure patterns under the effects of urbanization and urban migration. This scenario will create significant pressure to develop an age-friendly and accessible built environment to enhance health and functional capabilities among the older adult population in hill settlements. Subsequently, the chapter tries to gain a theoretical understanding of the functional capabilities of older adults from the lens of active ageing, ageing-in-place, active mobility and accessibility conditions in urban hill neighbourhoods and settlements through secondary literature resources. The chapter further highlights the accessibility and mobility challenges experienced by older adults residing in urban hill settlements caused by steep topographic conditions and planning and design constraints, frequently creating physical and psychological barriers. While stressing the need to promote active mobility opportunities and accessible public spaces, the chapter concludes by providing strategies and guidelines towards developing new or upgrading existing age-friendly neighbourhoods within hill settlements and thus working towards attaining the sustainable development goals of creating sustainable cities and communities.

The seventh chapter titled 'Assessment of the urban growth dynamics in Srinagar, Jammu and Kashmir, India', co-authored by Sneha et al. [39] investigates the urban growth dynamics and governance structures influencing Srinagar, an Indian Himalayan city experiencing rapid urbanization from 1930 to 2021. It unveils a complex interplay of historical, political, and governance factors shaping the city's evolution. The analysis reveals an exponential urban expansion primarily along the north–south axis, leaving significant underdevelopment in the south-eastern and eastern areas. Despite efforts like the Master Plans of 1971–1991 and 2001–2021, inadequate infrastructure persists, reflecting governance challenges. The emergence of Srinagar Smart City underscores a shift towards sustainable development. However, addressing spatial inequalities and enhancing socio-economic inclusivity remain critical. This research advocates for effective governance mechanisms, inclusive growth strategies, and innovative urban planning to foster a vibrant and resilient urban future for Srinagar. Such endeavours align with broader objectives of sustainability and equity, setting a precedent for urban development in the Indian Himalayan Region.

The eighth chapter titled 'Satellite Remote Sensing Techniques for Air Quality Monitoring in the Mid-Brahmaputra Valley of Northeast India', authored by Barman [40] examines the impact of air pollution on the Brahmaputra Valley. The study highlights that amidst the distinct climatic and physical features of these regions, the long-term analysis of atmospheric particles remains largely unexplored. Leveraging both ground-based in-situ particulate matter measurements and remote sensing observations, the chapter unveils a comprehensive examination of air particle dynamics across diverse seasons in the Mid-Brahmaputra Valley. As per World Health Organization (WHO) data, air pollution afflicts 99% of the global populace, with low- and middle-income countries bearing the greatest burden. The interconnectedness of air pollution with climate change adds complexity to the issue. While atmospheric particles emerge as major contributors to air pollution, research on them gains global importance. Despite sporadic studies on particulate matter and black carbon behaviour in various Indian locales, temporal patterns of air particles in north-eastern India remain largely uncharted. In regions lacking adequate ground monitoring resources, satellite data emerges as a crucial tool, offering insights into air quality and filling spatial data gaps. By documenting the spatial and temporal behaviour of atmospheric particles, this chapter sheds light on the complex relationship between air pollution and climate change, paving the way for future research directions and policy interventions.

The ninth chapter titled 'Spatio-Temporal Comparative Study of Aerosol Optical Depth, Aerosol Index, and Planetary Boundary Layer Height During and After COVID-19 Lockdown over North Eastern Region of India', authored by Saini and Lataye [41] analyses Aerosol Optical Depth (AOD), Aerosol Index (AI), and Planetary Boundary Layer Height (PBLH) in the context of India's COVID-19 lockdown and post-lockdown phases. MODIS Terra and OMI Aura datasets were utilized to investigate the interplay among human activity, air quality, and atmospheric conditions. Notably, the study uncovers an increase in AOD and AI levels during the second wave of COVID-19 lockdown in 2021, compared to the initial wave in 2020. Moreover, elevated levels persist during the post-lockdown phase in 2021, signalling a resurgence in anthropogenic and industrial activities, which contribute significantly to air quality deterioration. Further, the analysis focuses on the Northeast region of India, wherein positive AI values indicated the prevalence of absorbing aerosols such as dust, smoke, and biomass emissions. Furthermore, the impact of absorptive aerosols on PBLH leads to atmospheric warming. Intriguingly, contrary to the traditional inverse relationship between AOD and PBLH, this research reveals a simultaneous increase in both parameters over three years, especially during winter months, underscoring the complex dynamics of aerosol composition and atmospheric conditions.

The tenth chapter titled 'Geoinformatics for Climate Change Vulnerability Assessment in the Western Himalayan Cities', authored by Verma et al. [42] discusses the application of GIS, remote sensing, and GPS to assess climate change vulnerability in Western Himalayan cities. The proliferation of cities in Western Himalayas in an unplanned manner renders them highly susceptible to climate change due to their unique geographical and topographical characteristics. The compounding factors of rapid urbanization, population growth, and evolving urban land use patterns further exacerbate the vulnerability of cities in this region. Considering the peculiar issues of the hill regions, Geoinformatics, along with its components, offers spatially explicit applications for urban planning, land use management, land cover monitoring, urban transport planning, distribution of essential services, management of natural hazards such as landslides and urban flash floods, establishment of early warning systems, susceptibility mapping, municipal solid waste monitoring, collection, and landfill management for Western Himalayan cities. Additionally, this chapter addresses the limitations of Geoinformatics in assessing climate change vulnerability in these cities.

The eleventh chapter titled 'Assessing and Mapping Tourism Vulnerability using Remote Sensing and Spatial Analysis: A Case of Shimla District in Himachal Pradesh, India', authored by Manisha et al. [43] attempts to identify and evaluate the indicators influencing tourism vulnerability in Shimla District through spatial mapping and assessment. The study highlights the lack of previous scientific assessments of tourism vulnerability in Shimla District, emphasizing the importance of such analyses for effective risk reduction and sustainable tourism development. The chapter discusses the various factors contributing to tourism vulnerability, including natural hazards, climate change, economic fluctuations, and political disturbances. It also delves into the demographic and geographical characteristics of Shimla District, such as its population, climate, and topography. The research methodology adopted for this study is outlined, including data collection from online sources and expert opinions through questionnaires. The chapter details the classification of 18 indicators into Exposure, Sensitivity, and Adaptive Capacity criteria, which are essential for assessing and mapping tourism vulnerability. Thematic maps and cumulative criteria mapping are presented to visualize the spatial distribution of vulnerability indicators in Shimla District. Overall, the chapter sets the stage for understanding the complex interplay of factors influencing tourism vulnerability in Shimla and the need for sustainable tourism practices to mitigate risks and promote long-term resilience.

The twelfth chapter titled 'Safeguarding the Western Himalayas: Sustainable Disaster Management in the Hilly Regions of Himachal Pradesh, co-authored by Sharma et al. [44] focuses on the importance of sustainable disaster management in Himachal Pradesh, especially safeguarding the Western Himalayas. The chapter uses a rigorous and systematic methodology to showcase an understanding of the complexities in disaster risk assessment, utilizing advanced techniques such as Shannon's Entropy and the Fuzzy Technique. Additionally, GIS technology is utilized to identify vulnerable areas, enhancing the precision of the analysis. By performing an extensive literature review, the discussion contextualizes the concepts of disaster risk and vulnerability within the unique geographical and climatic environment of Himachal

Pradesh. The multifaceted vulnerability in the region is analysed through a detailed investigation of various disasters that occur throughout the state. Furthermore, it engages in a robust discussion, questioning the methodological foundations and extrapolating their potential scalability and broader applicability. Finally, it explores the sustainable disaster management, examining the current situation and advocating for the seamless integration of disaster risk reduction principles into development agendas to achieve resilient and sustainable outcomes.

The thirteenth chapter titled 'Issues, Options and Strategies for making Hill Habitat Inclusive, Safe, Resilient and Sustainable', authored by Gupta [45] emphasizes the unique challenges and opportunities associated with planning and development in hill areas. It begins by highlighting the distinctiveness of hill habitats compared to plain areas, emphasizing the necessity of specialized training and expertise for effective planning and development. The chapter discusses key issues faced by hill areas, including heritage conservation, accessibility, sustainable building design, and community involvement. Strategies for addressing these challenges are proposed, such as leveraging technology, integrating climate resilience, and formulating appropriate policy frameworks. Additionally, the chapter emphasizes the ethical considerations inherent in development activities in hill areas. Real-life examples are incorporated to illustrate successful approaches to heritage conservation, community engagement, and climate resilience. Overall, the chapter underscores the importance of adopting a holistic and context-specific approach to hill habitat planning and development.

In fourteen chapter titled 'Academic Contribution and Policy Perspectives: Geoinformatics for Sustainable Human Settlements in Hilly Regions', authored by Dahiya et al. [46] presents a discussion on the academic and policy contribution of the studies.

5 Concluding Remarks

The challenges and opportunities for human settlements and sustainable development in hilly regions are multifaceted and complex in the twenty-first century. Human settlements in these areas are intricately linked to environmental, socio-cultural, and economic dynamics, presenting both challenges and opportunities for sustainable development. The unique geomorphology, rugged topography, microclimatic variations, and susceptibility to natural hazards in hilly regions pose distinct challenges that necessitate the design, development and application of innovative approaches to urban planning, infrastructure development, and natural resource management. While the impacts of unsustainable development in hilly regions are escalating, the relevant efforts to address these issues within their fragile ecological context are imperative. However, the assessment of sustainability in environmentally sensitive hill regions remains a focus of limited research and academic attention. Moreover, the impacts of sustainability dimensions vary across different levels, further complicating the issue.

Geoinformatics emerges as a transformative solution to address the complexities of hilly terrain [47]. By integrating GIS, remote sensing, and spatial analysis has enabled stakeholders to gather, analyse, and visualize spatial data, providing valuable insights into the interactions between human activities and the environment. This technology offers a comprehensive framework for understanding, analysing, and managing the complex relationships between human and the environment in hilly regions. The application of geoinformatics in hilly regions extends to various domains, including land use planning, environmental monitoring, disaster risk management, and climate change vulnerability assessment. Through terrain analysis, natural resource management, and environmental monitoring, geoinformatics enables informed decision-making for sustainable development. In recent years, the application of geoinformatics has revolutionized the understanding and management of hilly landscapes, offering distinct approaches to address the challenges faced by human settlements in these regions. From terrain analysis to disaster risk management, geoinformatics offers a holistic approach to sustainable development in hilly regions. Therefore, the integration of geoinformatics into research and development endeavours in hilly regions is essential for promoting sustainable human settlements. By leveraging spatial data and advanced analytical techniques, stakeholders can make informed decisions to mitigate the adverse impacts of haphazard and unsustainable urbanization and safeguard the ecological integrity and cultural heritage of hilly regions.

References

1. Garg P, Kaur H (2020) Sustainability issues in context of Indian hill towns. In: Smart innovation, systems and technologies. Springer, pp 629–639. https://doi.org/10.1007/978-981-32-9868-2_53
2. NITI Aayog: Sustainable development in the Indian Himalayan region, https://www.niti.gov.in/sustainable-development-indian-himalayan-region. Accessed 1 Apr 2024
3. Department of Science and Technology (Government of India) (2010) National mission for sustaining the Himalayan eco-system
4. Independent Group of Scientists appointed by the United Nations Secretary-General (2019) Global sustainable development report 2019: the future is now—science for achieving sustainable development, New York
5. Wang Y, Wu N, Kunze C, Long R, Perlik M (2019) Drivers of change to mountain sustainability in the Hindu Kush Himalaya. In: The Hindu Kush Himalaya assessment: mountains, climate change, sustainability and people, pp 17–56. https://doi.org/10.1007/978-3-319-92288-1_2/TABLES/1
6. Nilsson C, Grelsson G (1995) The fragility of ecosystems: a review. J Appl Ecol 32:677. https://doi.org/10.2307/2404808
7. Pushplata (2000) Urban design matrix for hill towns
8. Akadiri PO, Chinyio EA, Olomolaiye PO (2012) Design of a sustainable building: a conceptual framework for implementing sustainability in the building sector. Buildings 2:126–152. https://doi.org/10.3390/BUILDINGS2020126
9. Lützkendorf T, Balouktsi M (2017) Assessing a sustainable urban development: typology of indicators and sources of information. Procedia Environ Sci 38:546–553. https://doi.org/10.1016/J.PROENV.2017.03.122

10. Dey J, Sakhre S, Gupta V, Vijay R, Pathak S, Biniwale R, Kumar R (2018) Geospatial assessment of tourism impact on land environment of Dehradun, Uttarakhand, India. Environ Monit Assess 190(4). https://doi.org/10.1007/s10661-018-6535-4
11. Mohan SR (2020) Post-tsunami reconstruction and panchayats: political economy barriers to effective implementation. In: Ghosh M (ed) Perception, design and ecology of the built environment: a focus on the global south. Springer, pp 135–148. https://doi.org/10.1007/978-3-030-25879-5
12. Thomas C, Vengattil M, Devasia J (2024) Landslides after heavy rain in India's Kerala kill 106, many still trapped. Reuters. https://www.reuters.com/world/india/several-feared-dead-after-landslides-indias-kerala-local-media-report-2024-07-30/
13. Sah NK, Pande RK (1987) Construction activity and environmental degradation in Almora town in the central Himalaya. Mt Res Dev 7:71. https://doi.org/10.2307/3673325
14. Chettry V (2022) Peri-urban area delineation and urban sprawl quantification in Thiruvananthapuram urban agglomeration, India, from 2001 to 2021 using geoinformatics. Appl Geomatics. https://doi.org/10.1007/s12518-022-00460-0
15. Chettry V (2023) A critical review of urban sprawl studies. J Geovis Spat Anal 7:28. https://doi.org/10.1007/s41651-023-00158-w
16. Persai P, Katiyar SK (2018) Development of information evaluation system for smart city planning using geoinformatics techniques. J Indian Soc Remote Sens 46:1881–1891. https://doi.org/10.1007/s12524-018-0844-5
17. Chettry V, Surawar M (2020) Urban sprawl assessment in Raipur and Bhubaneswar urban agglomerations from 1991 to 2018 using geoinformatics. Arab J Geosci 13:667. https://doi.org/10.1007/s12517-020-05693-0
18. Praharaj S, Solis P, Wentz EA (2023) Deploying geospatial visualization dashboards to combat the socio-economic impacts of covid-19. Environ Plann B: Urban Anal City Sci 50(5):1262–1279. https://doi.org/10.1177/23998083221142863
19. Praharaj S, Han H (2021) Human mobility impacts on the surging incidence of covid-19 in India. Geogr Res 60(1):18–28. https://doi.org/10.1111/1745-5871.12502
20. Ayambire RA, Amponsah O, Peprah C, Takyi SA (2019) A review of practices for sustaining urban and peri-urban agriculture: implications for land use planning in rapidly urbanising Ghanaian cities. Land Use Policy 84:260–277. https://doi.org/10.1016/j.landusepol.2019.03.004
21. Li J, Zhang H, Sun Z (2020) Spatiotemporal variations of land urbanization and socioeconomic benefits in a typical sample zone: a case study of the Beijing-Hangzhou grand canal. Appl Geogr 117:102187. https://doi.org/10.1016/j.apgeog.2020.102187
22. Gupta R, Sharma LK (2020) Efficacy of spatial land change modeler as a forecasting indicator for anthropogenic change dynamics over five decades: a case study of Shoolpaneshwar wildlife sanctuary, Gujarat India. Ecol Indic 112:106171. https://doi.org/10.1016/j.ecolind.2020.106171
23. Badar RN, Bahadure SP (2020) Assessing tourism sustainability in hill towns: case study of Shimla, India. J Mt Sci 17:2241–2261. https://doi.org/10.1007/s11629-019-5683-5
24. Chettry V, Manisha K (2022) Assessing and predicting urban growth patterns using ANN-MLP and CA model in Jammu urban agglomeration, India. In: Das B, Patgiri R, Bandyopadhyay S, Balas VE (eds) Modelling, simulation and optimization. Smart innovation, systems and technologies. Springer, pp 387–397. https://doi.org/10.1007/978-981-19-0836-1_30
25. Manisha K, Singh I, Chettry V (2023) Investigating and analyzing the causality amid tourism, environment, economy, energy consumption, and carbon emissions using Toda–Yamamoto approach for Himachal Pradesh, India. Springer Netherlands. https://doi.org/10.1007/s10668-023-04252-3
26. Manisha K, Chettry V (2024) Mapping cyclone and flood hazard vulnerability in Puri District, Odisha, India using geoinformatics. In: Devi NR, Ang S, Nawawi NM (eds) Sustainable resilient built environments: proceedings of SRBE 2022, India. Springer Nature, Singapore
27. Sharma K (2019) Urbanization induced land use-land cover changes in the Manipur valley and surrounding hills: a landscape metrics approach. In: Environmental change in the Himalayan

region. Springer International Publishing, pp 137–155. https://doi.org/10.1007/978-3-030-03362-0
28. Patowary S, Kumar A (2017) Model-based analysis of urban settlement process in eco-sensitive area of developing country: a study with special reference to hills of an Indian city. Environ Dev Sustain. https://doi.org/10.1007/s10668-017-9965-1
29. Patowary S, Sarma AK (2019) Projection of urban settlement in eco-sensitive hilly areas and its impact on peak runoff. Environ Dev Sustain. https://doi.org/10.1007/s10668-019-00453-x
30. Chhetri R, Kumar P, Pandey VP, Singh R, Pandey S (2020) Vulnerability assessment of water resources in hilly region of Nepal. Sustain Water Resour Manag 6:1–12. https://doi.org/10.1007/s40899-020-00391-x
31. Liu C, Cheng L, Li J, Lu X, Xu Y, Yang Q (2023) Trade-offs analysis of land use functions in a hilly-mountainous city of northwest Hubei Province: the interactive effects of urbanization and ecological construction. Habitat Int 131:102705. https://doi.org/10.1016/j.habitatint.2022.102705
32. Kapoor N, Jain M, Bansal VK (2019) GIS-based framework for local spatial planning in hill areas. Int Rev Spat Plann Sustain Dev 7:113–130. https://doi.org/10.14246/irspsd.7.4_113
33. Chettry V, Kaur H, Praharaj S, Dahiya B (2025) Introduction: geoinformatics for sustainable human settlements in hilly regions. In: Chettry V, Kaur H, Praharaj S, Dahiya B (eds) Geoinformatics for hilly regions: sustainable human settlement. Springer
34. Kaur H, Nawani A (2025) Sustainable urban development in hilly regions: concepts, evolution, and challenges. In: Chettry V, Kaur H, Praharaj S, Dahiya B (eds) Geoinformatics for hilly regions: sustainable human settlements. Springer
35. Sharma S, Saini A, Shrivastava B, Kumar A (2025) Land use planning in hill cities of India and abroad: a comparative study. In: Chettry V, Kaur H, Praharaj S, Dahiya B (eds) Geoinformatics for hilly regions: sustainable human settlements. Springer
36. Bahadure S, Kamble TM, Suthar ML (2025) Sustainable land-use planning in environmental sensitive hill city Gangtok, India. In: Chettry V, Kaur H, Praharaj S, Dahiya B (eds) Geoinformatics for hilly regions: sustainable human settlements. Springer
37. Gujar S, Kulkarni V, Maurya A, Bahale S (2025) The impact of culture, climate, topography, and geology on settlements in hilly areas: a case of Kandad, India. In: Chettry V, Kaur H, Praharaj S, Dahiya B (eds) Geoinformatics for hilly regions: sustainable human settlements. Springer
38. Rehal P, Chani PS, Atreya S, Nawani A, Kaur H (2025) Accessibility for senior citizens in hill settlements: challenges and opportunities for sustainable mobility. In: Chettry V, Kaur H, Praharaj S, Dahiya B (eds) Geoinformatics for hilly regions: sustainable human settlements. Springer
39. Sneha, Chettry V, Yadav A (2025) Assessment of urban growth dynamics in Srinagar (J&K), India. In: Chettry V, Kaur H, Praharaj S, Dahiya B (eds) Geoinformatics for hilly regions: sustainable human settlements. Springer
40. Barman N (2025) Satellite remote sensing techniques for air quality monitoring in the mid-Brahmaputra valley of Northeast India. In: Chettry V, Kaur H, Praharaj S, Dahiya B (eds) Geoinformatics for hilly regions: sustainable human settlements. Springer
41. Saini D, Lataye DH (2025) Spatio-temporal comparative study of aerosol optical depth, aerosol index, and planetary boundary layer height during and after covid-19 lockdown over north eastern region of India. In: Chettry V, Kaur H, Praharaj S, Dahiya B (eds) Geoinformatics for hilly regions: sustainable human settlements. Springer
42. Verma R, Kaur H, Dwivedi A (2025) Geoinformatics for climate change vulnerability assessment in the western Himalayan cities. In: Chettry V, Kaur H, Praharaj S, Dahiya B (eds) Geoinformatics for hilly regions: sustainable human settlements. Springer
43. Manisha K, Singh I, Jain M (2025) Assessing and mapping tourism vulnerability using remote sensing and spatial analysis techniques: a case of Shimla district in Himachal Pradesh, India. In: Chettry V, Kaur H, Praharaj S, Dahiya B (eds) Geoinformatics for hilly regions: sustainable human settlements. Springer

44. Sharma S, Sharma S, Sharma S (2025) Safeguarding the western himalayas: sustainable disaster management in the hilly regions of Himachal Pradesh. In: Chettry V, Kaur H, Praharaj S, Dahiya B (eds) Geoinformatics for hilly regions: sustainable human settlements. Springer
45. Gupta JK (2025) Issues, options and strategies for making hill habitat inclusive, safe, resilient and sustainable. In: Chettry V, Kaur H, Praharaj S, Dahiya B (eds) Geoinformatics for hilly regions: sustainable human settlements. Springer
46. Dahiya B, Chettry V, Kaur H, Praharaj S (2025) Academic contribution and policy perspectives: geoinformatics for sustainable human settlements in hilly regions. In: Chettry V, Kaur H, Praharaj S, Dahiya B (eds) Geoinformatics for hilly regions: sustainable human settlement. Springer
47. United Nations (2021) Geospatial strategy for the United Nations. https://www.un.org/geospatial/sites/www.un.org.geospatial/files/20210330_geospatial_strategy_for_the_united_nations-book.pdf

Sustainable Urban Development in Hilly Regions: Concepts, Evolution, and Challenges

Harsimran Kaur and Akhil Nawani

1 Introduction

Urbanization is an unstoppable global phenomenon, with more than half the world's population residing in urban areas. This rapid urban growth poses various challenges, including increased resource demand, environmental degradation, and social inequalities. To address these issues, sustainable urban development has emerged as a guiding framework for planning and managing cities. Sustainable urban development is an approach that aims to establish environmentally, socially, and economically viable cities and communities over the long term. It acknowledges the need to balance urban growth and development with the preservation of natural resources, the well-being of residents, and the reduction of negative environmental consequences. However, the environmental susceptibility and topographical complexity of hilly regions introduce unique challenges and opportunities that require specialized strategies for sustainable urban development.

Sustainable urban development focuses on fundamental principles for fostering resilient and harmonious cities. One of these pivotal principles is Environmental Conservation. Cities adopting sustainable development practices prioritize preserving natural resources like clean water, air quality, and green spaces. Efforts are made to reduce pollution and protect local ecosystems. Another foundational pillar of sustainable urban development is Economic Viability. This principle aims to create a robust and diverse economy by fostering innovation, supporting local businesses, and attracting investments. It seeks to reduce poverty and inequality by providing job opportunities and access to services for all residents. Social Inclusivity stands as an

H. Kaur (✉) · A. Nawani
Department of Architecture, Planning and Design, Indian Institute of Technology (BHU) Varanasi, Varanasi, Uttar Pradesh, India
e-mail: harsimran.apd@iitbhu.ac.in

A. Nawani
e-mail: akhil.apd@iitbhu.ac.in

V. Chettry et al. (eds.), *Geoinformatics for Hilly Regions*, Advances in 21st Century Human Settlements, https://doi.org/10.1007/978-981-95-0156-4_2

integral component of this paradigm, which aims to ensure that all residents, regardless of their socio-economic status, have access to essential services such as housing, education, healthcare, and public transportation. It promotes cultural diversity and social cohesion. Efficient infrastructure remains another linchpin of sustainable cities, prioritizing efficient and well-planned infrastructure, including public transportation, waste management, and energy systems. Smart urban planning reduces congestion, energy consumption, and greenhouse gas emissions. Resilience serves as a cornerstone of sustainable urban development, emphasizing meticulous planning for the impacts of climate change and natural disasters, striving to make cities more resistant and adaptive. This includes strategies for flood control, disaster preparedness, and sustainable building practices.

Shifting our attention toward sustainable urban development in hilly regions, it becomes evident that these places pose distinct challenges and prospects. Hilly terrain necessitates meticulous planning and engineering measures to mitigate the adverse effects of soil erosion, landslides, and other environmental hazards. The integration of land use planning that demonstrates reverence for the inherent characteristics of the natural environment is vital for achieving sustainable development in hilly regions. The hilly topography presents distinct obstacles that also impact the field of transportation. The challenging topography frequently requires implementing effective and environmentally sustainable solutions for public transit systems, such as cable cars, funiculars, or narrow-gauge railways. Hilly terrains present favorable conditions for establishing green corridors, parks, and sustainable agricultural practices. Preserving trees and natural vegetation is crucial in mitigating soil erosion and sustaining biodiversity.

Moreover, hilly regions often boast access to renewable energy sources like hydroelectric power. Sustainable urban development can harness these resources for clean energy production. Cultural Heritage and Historic Settlements are intrinsic to many hilly regions, demanding preservation and integration into urban planning promoting tourism and cultural identity. Given the susceptibility of hilly regions to landslides and other natural disasters, sustainable urban development should include robust disaster preparedness and response plans.

In conclusion, sustainable urban development is essential in both conventional urban settings and hilly regions. While the fundamental principles remain the same, addressing the unique challenges of hilly terrain requires specific strategies and considerations. Cities in these areas truly thrive through careful planning and a commitment to preserving their natural beauty and resilience.

2 Sustainable Urban Development: Concepts and Principles

2.1 Sustainable Urban Development and Three Pillars

Sustainable urban development is a multifaceted concept that seeks to harmonize urban growth's economic, social, and environmental dimensions. It aspires to create resilient, inclusive, and environmentally responsible cities. Fundamental principles include resource efficiency, social equity, economic viability, and environmental stewardship. Sustainability in urban development is often depicted through three interconnected pillars.

Environmental Sustainability—Management of natural resources • Reduction of greenhouse gas emissions • Preservation of biodiversity.

Social Sustainability—Equitable access to housing, healthcare, and education • Social cohesion and community engagement • Inclusivity for all demographic groups.

Economic Sustainability—Diversification of the economy • Job creation and income generation • Efficient resource allocation.

2.2 Models of Hill Area Development

Globally, there have been a couple of approaches i.e., the European and Japanese models which have been used extensively and made operational for hill area development. One other approach has been lauded for its unique tactic through high prioritization of environmental preservation. These models differ in their approach, intent and content related to the development of hill areas while conserving and valuing the fragility and ecology of such areas. European countries have adopted the principle and strategy of hill area development by involving, including, and using sound practices of land use planning duly supported by efficient land tenure system; development of hydro-electric power as a significant source of alternate energy; preventing the destruction of forest cover and adopting the development of planned and dispersed tourism. All these factors have been integrated to conserve, preserve, and develop hill areas without destruction. In the Japanese model of hill area development, hills have been made areas of low-density development, housing only 2% of the population while occupying 75% of land areas.

In contrast, the remaining 98% of the population is housed in 25% of the land area. In the Japanese model, hill areas have been mainly used for the conservation of resources, forestry, generating hydro-electric power, and promoting tourism. In contrast, plain areas have been used for urbanization and human living, housing large-scale economic activities, including industry, trade, and commerce, despite the adverse impact of acute congestion and pollution. These models have relevance, advantages, limitations, importance, and contexts. They can be considered for adoption in the Indian context, with appropriate modifications, to suit the specific needs and prevailing physical, cultural and socio-economic conditions in different hill states.

The models of hill area development from European, Swiss, Japanese, and Bhutanese contexts can certainly be adapted and applied to the Indian and broader South Asian context. However, it is essential to consider the region's unique characteristics, cultural aspects, and challenges. Here are how these models can be customized and utilized in the Indian and South Asian context.

The European Model

Sustainable Tourism and Cultural Preservation. Emphasize sustainable tourism practices while preserving the rich cultural heritage of the Indian and South Asian hill areas. This could involve preserving historic sites, indigenous arts and crafts, and traditional festivals.

Integrated Transportation. Develop efficient transport infrastructure tailored to the local geography. This may include improved road networks, mountain railways, and cable cars for convenient access.

Stringent Zoning Regulations. Enforce zoning regulations that respect the ecological and aesthetic integrity of the region, considering local architectural styles and the preservation of natural landscapes.

Community Engagement. Actively involve local communities in decision-making and development planning to ensure their unique perspectives and interests are considered.

The Japanese Model

Cultural and Natural Preservation. Celebrate the unique cultural and natural heritage of the Indian and South Asian hill areas by preserving historic sites, gardens, and indigenous traditions.

Efficient Infrastructure. Develop and maintain efficient road networks and rail systems to facilitate access.

Sustainable Agriculture and Farm Tourism. Encourage sustainable agriculture and offer farm tourism experiences, allowing visitors to experience and participate in local farming practices.

Local Community Involvement. Engage local communities in tourism development, ensuring that their perspectives and needs are integrated into planning and decision-making.

The Bhutanese Model

Gross National Happiness: Consider a focus on well-being and happiness over economic growth, like Bhutan's Gross National Happiness approach, in the context of the Indian and South Asian hill areas.

Low-Impact Tourism: Regulate visitor numbers and emphasize sustainable and low-impact tourism practices to protect the environment and culture.

Community-Based Development: Encourage community-based tourism and ensure that local communities benefit from tourism revenues.

Environmental Conservation. Prioritize environmental conservation and the protection of biodiversity in the hill areas.

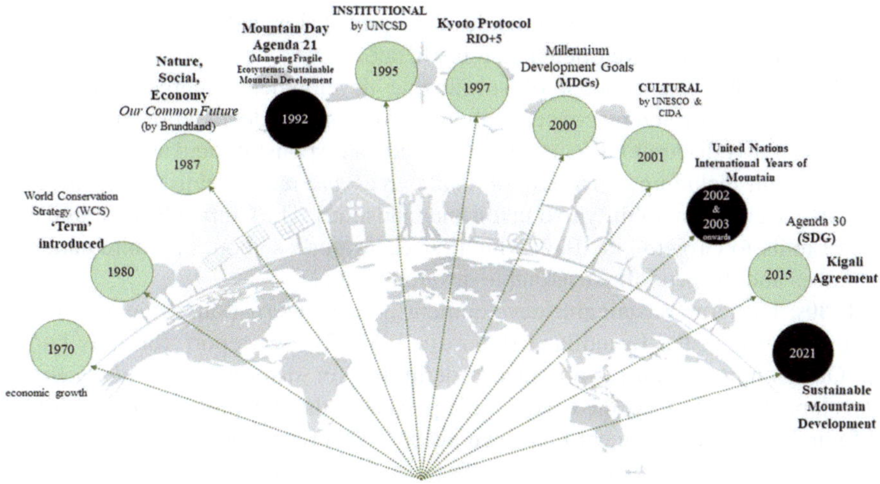

Fig. 1 Sustainable urban development milestones over the years. *Source* Author

In the Indian and South Asian context, adaptation is crucial, and any development models should consider the diverse cultures, ecological challenges, and local needs. Engaging with local stakeholders, respecting traditional practices, and preserving the natural environment are essential components of responsible hill area development in the region.

3 Evolution of Sustainable Urban Development

3.1 Historical Perspectives

The concept of sustainable urban development has evolved over time, driven by various global and regional factors (Fig. 1). Early urban planning efforts primarily focused on hygiene and infrastructure. It was not until the mid-twentieth century that environmental concerns gained prominence, leading to the birth of sustainable development.

3.2 Key Milestones in Sustainable Urban Development

Sustainable urban development, in general, and specific to hill regions, has become an increasingly important topic as cities worldwide grapple with population growth, climate change, and resource constraints. Here is a timeline highlighting key events and developments in sustainable hillside development alongside the broader timeline for sustainable urban development and resilience:

1960s–1970s: The concept of sustainable development emerged during this period, primarily driven by concerns about environmental degradation and resource depletion. Early discussions focused on the need for responsible land use and resource management, including considering hillside areas and recognizing their ecological importance and vulnerability to erosion and landslides. The Club of Rome published "The Limits to Growth" in 1972, which raises awareness about the finite nature of resources and the need for sustainable development [1].

1980s: The term "sustainable development" gained international recognition with the publication of the Brundtland Report in 1987 [2]. The report, formally known as "Our Common Future," provided a definition for *sustainable development,* as "development that meets the needs of the present without compromising the ability of future generations to meet their own needs." This report played a pivotal role in shaping the sustainable development discourse. The Thistle Landslide in 1983 in Utah, USA, served as a wake-up call about the dangers of poorly planned hillside development. This disaster led to increased scrutiny of hillside development practices. Studies on soil erosion and slope stability gain prominence in geotechnical engineering, contributing to safer development on hillsides.

1990s: Sustainable urban development became a more prominent concept during this decade. The United Nations Conference on Environment and Development (UNCED), also known as the Earth Summit, was held in 1992 in Rio de Janeiro. It resulted in adopting Agenda 21 (at the local level), which included a specific chapter on sustainable urbanization. This marked a significant step in recognizing the importance of cities in the context of sustainable development. The Kyoto Protocol (1997) was adopted, marking a significant step in addressing climate change. It emphasizes the importance of urban planning and energy efficiency. Cities like Curitiba, Brazil, implement innovative urban planning and public transportation systems, setting an example for sustainable urban development. Researchers and planners have begun to emphasize the importance of preserving natural vegetation on hillsides to prevent erosion and enhance stability. Regulations and guidelines for hillside development emerge in various countries and regions, including California, where local ordinances are implemented to manage development in hillside areas.

2000s: The United Nations' adoption of the Millennium Development Goals (MDGs) in 2000 included goals related to urban sustainability, such as Goal 7, which aimed to ensure environmental sustainability and improve slum conditions. The United Nations Habitat Agenda, adopted at the Habitat II conference 1996, calls for sustainable urban development and adequate housing and infrastructure in cities. Sustainable building practices such as green building certification systems incorporate guidelines for hillside construction, including erosion control measures and sustainable landscaping. The United Nations' focus on sustainable development begins to focus on hilly and mountainous regions as critical areas for conservation and responsible development.

2010s: In 2015, the Sustainable Development Goals (SDGs) were established, with Goal 11 explicitly focusing on making cities and human settlements inclusive, safe, resilient, and sustainable. The Sendai Framework for Disaster Risk Reduction (2015) emphasizes the importance of resilience in natural disasters, recognizing that urban areas are increasingly vulnerable. Climate change and its impacts, including increased rainfall and landslides in hilly areas, bring additional attention to the need for sustainable hillside development. Advancements in technology, such as Geographic Information Systems (GIS) and remote sensing, were vital in assessing and mitigating the risks associated with hillside development.

2020s: Cities worldwide continue implementing sustainable urban development strategies, emphasizing green infrastructure, public transportation, and affordable housing. The COVID-19 pandemic prompts cities to rethink urban planning to ensure public health, accessibility, and resilience. Climate adaptation and mitigation strategies become a central focus as cities work to reduce greenhouse gas emissions and prepare for the impacts of climate change. Sustainable hillside development practices continue to evolve, strongly emphasizing minimizing environmental impact, preserving biodiversity, and ensuring community safety. Integrated land use planning and zoning codes increasingly consider the unique challenges and opportunities of hillside areas in urban and regional development plans. Awareness of the importance of preserving hillside ecosystems for carbon sequestration and climate resilience grows.

The timeline reflects the evolution of sustainable urban development and resilience as global priorities. These concepts have evolved from early environmental concerns to integrated approaches that consider economic, social, and environmental dimensions, with a strong emphasis on addressing climate change and building resilient communities. The timeline also highlights the persistent challenges and potential for cities to achieve sustainability and resilience in a dynamic and evolving global context.

4 Challenges in Sustainable Urban Development in Hilly Regions

Hilly regions present unique sustainability challenges, encompassing geographical, environmental, social, and economic aspects (Fig. 2). These challenges include geographical constraints such as steep slopes, unstable terrain, and limited space for development, necessitating innovative solutions in sustainable urban development planning [3]. Moreover, hilly areas are often ecologically sensitive, requiring special attention to preserve biodiversity and prevent soil erosion [4]. Additionally, effective water management in hilly regions is crucial to prevent the risk of landslides and floods. In these regions, marginalized communities may encounter limited access to essential services and economic opportunities, making social equity a central challenge in sustainable urban development efforts.

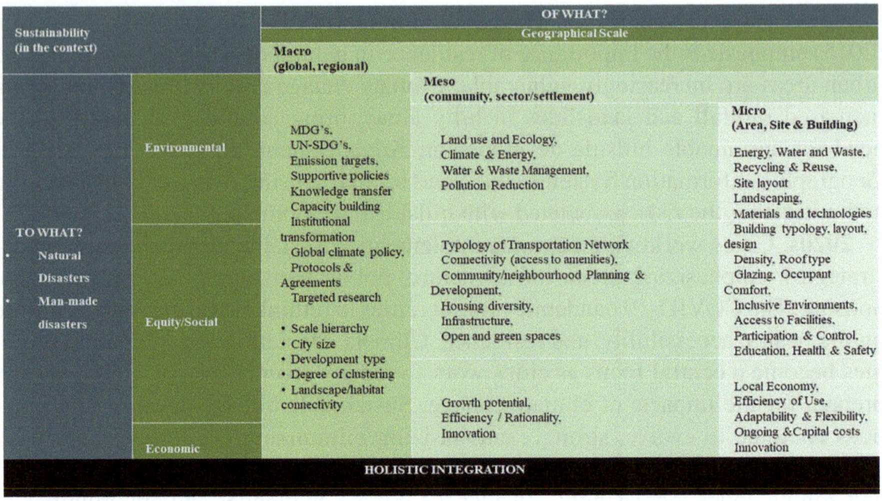

Fig. 2 Sustainability challenges at various levels. *Source* Evolved from [5]

These challenges can be understood at different levels, including the building, settlement, community, and global levels. Here are specific sustainability issues at each of these levels in hilly areas:

4.1 Building Level

Hillside Construction Practices: Building on steep slopes can lead to soil erosion, landslides, and unstable foundations. Sustainable building practices in hilly areas involve appropriate site analysis, slope stabilization, and erosion control measures [5].

Energy Efficiency and Heating: In colder hilly regions, buildings may require more energy for heating. Sustainable building practices include proper insulation, passive solar design, and the use of energy-efficient heating systems to reduce energy consumption.

Water Management: Hilly areas may have unique challenges related to water supply and drainage. Sustainable building practices involve rainwater harvesting, efficient water fixtures, and strategies to manage stormwater runoff and prevent erosion.

Natural Materials: Sustainable building materials sourced from the local environment can reduce transportation-related emissions and support the local economy. However, overharvesting can lead to environmental degradation, so responsible sourcing is crucial.

Avalanche and Landslide Risk: In regions prone to avalanches and landslides, buildings must be designed to mitigate these risks. This may involve avalanche barriers, rockfall protection, and proper foundation design.

4.2 Settlement Level

Land Use Planning: Sustainable land use planning in hilly areas involves minimizing urban sprawl and preserving green spaces and natural habitats. Developing comprehensive zoning regulations and building codes that consider the unique topography of hilly terrain is important [6].

Transportation: Steep and winding roads in hilly regions can lead to higher energy consumption and increased accident risks. Sustainable transportation planning includes the development of efficient road networks, public transit systems, and pedestrian-friendly infrastructure.

Disaster Preparedness: Settlements in hilly areas must have robust disaster preparedness plans and early warning systems in place to respond to natural hazards such as landslides, floods, and earthquakes [7].

Erosion Control: Sustainable settlement planning should prioritize erosion control measures, including reforestation, terracing, and the use of retaining walls to prevent soil erosion and landslides.

Access to Services: Ensuring that residents in hilly areas have equitable access to healthcare, education, and emergency services is crucial, especially given the challenging terrain.

4.3 Global Level

Climate Change Mitigation: Addressing climate change is a global priority. Sustainable practices in hilly regions can contribute to reducing greenhouse gas emissions through responsible land use, energy efficiency, and afforestation.

Biodiversity Conservation: Hilly regions often contain unique and diverse ecosystems. Global efforts to protect biodiversity should include the preservation of these fragile environments.

International Cooperation: Sustainable development in hilly areas may require cross-border cooperation, especially in regions with shared mountain ranges. Collaborative efforts can address transboundary issues like water resource management and conservation.

Disaster Risk Reduction: Global efforts in disaster risk reduction can help hilly regions prepare for and respond to natural disasters, minimizing the impact on both local communities and the global economy.

Economic Development: Sustainable economic development in hilly areas can reduce poverty and migration to urban centers, contributing to global economic stability and social equity.

Hilly regions face unique sustainability challenges, including geographical constraints and environmental vulnerabilities. Geoinformatics technology offers solutions for mapping terrain characteristics, assessing environmental risks, and optimizing land use planning. The application of remote sensing and GIS in addressing these challenges, emphasizing the need for interdisciplinary collaboration and community involvement.

5 Application of GIS and Remote Sensing Technologies

In light of the difficulties encountered by hilly regions, the implementation of GIS and remote sensing technology arises as a potentially effective strategy. By facilitating the acquisition of spatial data spanning extensive regions, remote sensing permits the detection and surveillance of environmental variables including terrain attributes, vegetation vitality, and land cover. Integration and analysis of spatial data are facilitated by GIS technology, allowing for more informed decisions to be made regarding urban planning and resource management. Through the utilization of remote sensing and GIS, policymakers and planners can acquire significant knowledge regarding the intricacies of hilly terrains. This enables them to devise precise interventions that promote sustainable urban development.

The utilization of GIS technology in conjunction with remote sensing permits the execution of focused mitigation tactics to promote sustainable urban development in hilly areas. Possible such strategies comprise.

Identification of High-Risk Areas: By utilizing remote sensing data to identify regions susceptible to soil erosion, landslides, and other environmental hazards, preventative risk mitigation measures can be implemented.

Land Use Planning: GIS-based land use planning enables the allocation of land for diverse objectives, while also considering environmental sensitivity and geographical limitations. [8]

Watershed Management: The implementation of remote sensing techniques can facilitate the delineation and monitoring of watersheds, thereby promoting efficient management of water resources and the prevention of floods.

Infrastructure Planning: To improve the resilience and sustainability of hilly regions, GIS technology facilitates the planning and design of infrastructure projects, including roads, drainage systems, and green spaces.

6 Conclusion

Sustainable urban development in hilly regions requires tailored strategies that leverage geoinformatics technology. By integrating remote sensing and GIS into planning and decision-making processes, cities can effectively address environmental, social, and economic challenges. International models offer valuable insights, but adaptation to local conditions is essential for success. Moving forward, interdisciplinary cooperation and community engagement will be key to achieving sustainability goals in hilly regions.

References

1. Meadows DH et al (1982) The limits to growth. A Potomac Associates Book
2. UNWCED Homepage (1987) Brundtland report. https://sswm.info/sites/default/files/reference_attachments/UNWCED1987BrundtlandReport.pdf. Accessed 20 Feb 2014
3. Pushplata, Kumar A (2012) Building regulations: a means of ensuring sustainable development in hill towns. Asian J Environ Res Dev 7A(1A):553–560
4. Sah N et al (2018) Hill slope instability of Nainital City, Kumaun Lesser Himalaya, Uttarakhand, India. J Rock Mech Geotechn Eng 10(2):280–289
5. Garg P, Kaur H (2019) Sustainability issues in context of Indian hill towns. In: Proceedings of the international conference on sustainability in energy and buildings, pp 629–639. Springer, Singapore
6. Kumar A, Pushplata (2013) Building regulations for environmental protection in Indian hill towns. Int J Sustain Built Environ 2(2):224–231
7. Prashar N, Lakra HS, Shaw R, Kaur H (2023) Urban flood resilience: a comprehensive review of Assessment methods, tools, and techniques to manage disaster. Prog Disaster Sci 20:1–17
8. Kaur H, Garg P (2023) Urban sustainability assessment tool for hillside planning, design and development. J Urban Plann Dev

Land Use Planning in Hill Cities of India and Abroad: A Comparative Study

Swasti Sharma, Ashutosh Saini, Bhavna Shrivastava, and Ashwani Kumar

1 Introduction

Land is considered the most essential resource and is used by humans for various purposes like agriculture, mining, constructing houses, shelters, roads, and setting up industries. But with the passage of time, the use of the land has increased due to the rise in the population and high demand of the people which resulted in the rapid development of the land mainly due to various construction activities [1].

In India, the concept of land use planning came when the requirement for the planned towns and cities was felt. Earlier as the population was not large enough, so the planning was carried out on the basis of locational preferences, class and caste system preferences, social customs, etc. [2, 3]. The land uses are now generally allocated on the basis of predictions on the economy, society, and technology made by various planners based on their knowledge [4]. There is an increasing pressure on land as a result of high demand and ineffective land use planning and management by the state has a negative impact on the environment and society [5]. Population growth which leads to haphazard urban development. The issue of unprecedented growth is more pronounced in the hill towns of India. The problem of the unplanned

S. Sharma
Amity School of Architecture and Planning, Amity University, Punjab, India
e-mail: 2020rar9574@mnit.ac.in

A. Saini
School of Architecture and Design, Manipal University, Jaipur, India

B. Shrivastava
Department of Architecture and Planning, Malaviya National Institute of Technology Jaipur, Jaipur, India
e-mail: bhavna.arch@mnit.ac.in

A. Kumar (✉)
Department of Architecture, National Institute of Technology, Hamirpur, India
e-mail: ashwanik@nith.ac.in

V. Chettry et al. (eds.), *Geoinformatics for Hilly Regions*, Advances in 21st Century Human Settlements, https://doi.org/10.1007/978-981-95-0156-4_3

development in the hills is due to the inappropriate land use which has resulted in various challenges in hill towns. Because there are not enough spatial planning methods, the land in the hilly areas is distributed without appropriate surveys, such as topographical maps showing the direction of the slope, the orientation of the sun, the type of flora, and the soil [6]. Unplanned development often leads to the emergence of natural disasters as one of its consequential challenges. SDG 11 encompasses the objective of enhancing the resilience of cities and human settlements to mitigate the impacts of natural disasters and various other challenges. The growth in the hills and the allocation of land use needs to be monitored and planned in a way that retains the unique and traditional touch and overcomes the challenges related to the hills.

2 Methodology

In order to comprehend the existing land use planning techniques in hill cities in India and abroad, systematic research has been conducted. The emphasis is on research that aids in analyzing the existing land use planning process in the hill towns. Using a combination of case studies, literature reviews, on-site observations, and qualitative data analysis, the study used a mixed-methods research strategy. The detail of the the specific methodologies used for data analysis is described as follows.

Review of the Literature and Database Search.

Methodology: Utilizing databases like Web of Knowledge and SCOPUS, as well as government publications.

Detail: To obtain pertinent literature and official papers, systematically search for the terms "land use," "India," "Nepal," "Bhutan," "the United States," and "land use planning."

Case Studies:

Approach: The study has focused on the best-case studies for land use planning in Indian hills, including New Tehri, Uttarakhand; Lavasa, Maharashtra; and Bilaspur, Himachal Pradesh. Similarly, the land use planning procedures used in the hill cities of Thimphu (Bhutan), Kathmandu (Nepal), Santa Clarita, Anchorage, and Boone County (United States) are examined in order to understand the land use planning processes followed in these hill cities.

Detail: In-depth examination of Indian hill towns for land use planning procedures through site visits, observations, and photographic record-keeping. For the international hill towns the detail of the land use planning process is gathered through their official sites or documents.

Observations Made On-Site:

Methodology: Visit Indian hill cities to study and document land use practices.

Detail: Using photos to record visual information for a more thorough analysis of the land use and urban character of these places. As the records or documentation of the land use planning in Bilaspur town were missing, the data was gathered by consulting various district officials and conducting a survey of individuals who have resided there for the previous 30 to 50 years.

Comparing Hill Cities in India with Hill cities in Abroad:

Methodology: Comparing land use planning strategies using qualitative data.

Details: Comparing and evaluating the land use planning procedures of hill towns in India (such New Tehri, Lavasa, and Bilaspur) and other countries (like Thimphu, Kathmandu, Santa Clarita, Anchorage, and Boone County) through a methodical qualitative data analysis. These case studies help to better understand the steps that must be taken for land use planning and bring out the various lacunas of planning in the Indian hills that guide India toward the better and planned development of its fragile areas.

3 Land Use Planning: A Literature Review

Land use planning is defined as an "administrative and statutory activity that seeks to regulate and order land utilization in an efficient and appropriate manner, thus avoiding land use conflicts; land use planning could control and determine future land use development to a large extent in planned areas" [7, 8]. Land is a limited and valuable resource, and governments strive to create policies, laws, plans, and strategies to maximize its use [5]. However, inefficiencies in land management abound, not only in India but throughout the world; the unplanned land uses and violations of land use regulations also cause various challenges for the environment and society. Numerous instances of widespread unplanned land uses can be found throughout India, such as Unauthorized construction on hill slopes, Indiscriminate land use conversion in peri-urban areas, growth of unplanned housing in cities, Urban sprawl, etc. [5].

These problems are particularly noticeable in the hills because of its geography, terrain, slope, climate, and susceptibility to natural hazards, which make land use planning and development there difficult [9]. Although master plans in India can control land development to some extent, they do not provide a scientific basis for allocating land use. It is vital to document land use planning in order to understand how land use in a specific area is determined. However, the existing literature contains relatively limited details about the procedure, such as statistical evidence or land use planning analysis. The allocation of different land uses is based on the requirements for the next twenty years. For example, in the master plan of Ghumarvien (a hill town) in Bilaspur district of Himachal Pradesh, various aspects were taken into account when making the proposal, including carrying capacity based on population, existing land use in the area, land potential, compliance of land use in the area's environment, threshold population for amenities, locational attributes of amenities, site conditions, and a feasible distance between work areas and residential areas [10]. However, the literature does not describe any process, i.e., how land use planning is conducted in the hills of India. The development plans or master plans for the hills do not include information regarding the creation of land use maps. The plans only indicate the projected population for the next 20 years and the proposed land uses, such as residential, commercial, industrial, etc., but neither the process nor the justification for allocating a particular land use to a particular parcel is mentioned. Also, the factors that are taken into account for the allocation of various land uses are given only for a few towns. However, it's a matter of concern as the hills are facing a tremendous increase in development activities and there are no set criteria for the allocation process of the land uses. These hill towns require a framework for the land use allocation process to provide the land with the best use.

4 Study of Existing Land Use Planning in Indian Hill Towns

The development of the hilly terrain depends on land use planning, which calls for a specific land use plan that addresses the socioeconomic, physical, cultural, and environmental challenge [11]. To understand the issues related to land use planning, a detailed case study of Bilaspur, New Tehri and Lavasa city is further carried out in the following section. The planning of these hill towns is done after the Independence. These towns (Bilaspur, Lavasa and New Tehri) were planned and developed as a part of rehabilitation efforts due to the construction of dams. A detailed discussion to understand the efforts of the Indian government in the domain of physical planning of these new hill towns are as follows.

4.1 Bilaspur

Bilaspur is a district in Himachal Pradesh and is locally known as 'Mini Chandigarh of Hills'. The establishment of the New Bilaspur Town is due to the Bhakra Nangal Dam project [12].

Land Use Planning of Bilaspur Town: The development plan for Bilaspur (2005) for the year 2021, mentions the allocation of various land uses is carried out based on analysis of various surveys, and public aspirations UDPFI guidelines issued by the Government of India. Factors that are considered for the projection of land use for the year 2021 are developable land in each sector, surrounding land uses, threshold population, locational attributes of amenities, site characteristics, workplace relationships and land values (Fig. 1). Despite practices for land use in Bilaspur such as surveys and prioritization of the factors for land use allocation, unauthorized and unplanned construction activities have been noticed in the district over time due to poor planning policies and rehabilitation practices in New Bilaspur Town [13].

The literature that describes the basis for land use allocation during 1960 is not available. Therefore, to reinvestigate the reason for unplanned development, a structured interview was conducted with various town planning officials of Himachal Pradesh to understand the basis for land use allocation in Bilaspur during the 1960s.

Fig. 1 Existing development in Bilaspur. *Source* Author, 2021

As per the experts, the planning was carried out in sectors based on the population projection and requirement of the area but it lacked enforcement, formulation, and regulation of various bye-laws, due to which land owners covered maximum area for the construction activity which resulted in unplanned development. Also, the plots or areas that were designated for residential purposes were converted to commercial purposes which resulted in the change of land use due to inefficient regulations and framework for New Bilaspur town. The lack of planning practices for land use in Bilaspur led to encroachments in the town.

Therefore, the planning for Bilaspur town proved to be a major challenge due to inefficient land use allocation and enforcement for the New Bilaspur Town. The land uses were not designated on the basis of scientific criteria such as model application or tools or techniques. For the planning of land uses, the new Bilaspur town was divided into sectors and various land uses were specified for different sectors. The regulation and enforcement of building bye-laws, ground coverage, and FAR were also not incorporated in the designs for various construction activities which lead to encroachments and inappropriate land uses in the town.

4.2 Lavasa

Lavasa is known as the first planned hill city in the state of Maharashtra. The master plan of Lavasa is based on the vision of "People living in harmony with nature" and it takes into term the principle of "New Urbanism", i.e., to live, play, work, and learn in harmony with nature.

Land use planning of Lavasa: The main intent of the land use planning in Lavasa was to plan in such a way that the hill city offers high-quality facilities and also protects the environment by utilizing the natural assets of the place [14]. The concept of walkability to reduce the pressure on transportation is also adopted in the master plan. The town center having the major public activities is in the center of each town and these workplaces are further surrounded by flats for the residents who work in these workplaces. Hence the distance from the workplace is reduced reducing the transportation cost as well as increasing the walkability in the areas. As the planning is based on the transect model principle, therefore the density of the residential area decreases as we move away from the town center. The Town center is located in the valley, thus the density of the residential area is less on the hilltops or the slopes [15].

The Lavasa project aimed to reach its heights by introducing good planning practices and designs to the site but due to several reasons, success didn't reach its way. Initially, the Lavasa project captivated people's interest, but it quickly devolved into a nightmare for investors that invested their life savings into this project. For the land use allocation, various surveys and analyses such as slope analysis, contour analysis, aspect analysis, and analysis based on hydrological information were performed to allocate the best use of land in the city. But the project has gone through challenges such as misuse of building bye-laws (construction of G + 6 buildings), environment degradation (cutting of slopes and deforestation), inappropriate conversion of land use, contamination of natural water sources, stone quarrying, etc. Initial plan approval has attracted a lot of investors throughout the world as the planning agency and the master plan of Lavasa claimed it to be a sustainable planned hill city with appropriate land uses. However, the evaluation committee concluded that the project could not achieve sustainability due to inappropriate land uses and the misuse of building bye-laws, and thus the project was stopped [16].

4.3 New Tehri

In February 2000, New Tehri received the PM's Award for Excellence in Urban Planning and Design. The New Tehri town was built to house the residents of Old Tehri who had been displaced by the construction of the Tehri Dam. The town covers an area of 80 hectares (200acres) with natural terrace ridge (eastern and south-east slope) valleys and has a dome-shaped form that reaches out down to a flat terrace. The population of the town in the year 2011 was 24,014.

Land Use Planning in New Tehri: Land use has been planned in layers. The town has a mix-land-use pattern, with the majority of educational and institutional structures on the elevated terraces and main administrative and public functions on the fairly low terraces, while residential development is concentrated on the mid terraces. The township was designed as a series of cascading clusters of buildings (Fig. 2). Vast

Fig. 2 Scenic view of Tehri Dam. *Source* Author, 2019

open areas in the town center were designed as public spaces for meetups, reunions, and leisure activities. The town is planned in such a way that it promotes walkability by providing the maximum facilities within 400m of the dwelling units. The author has mentioned accessibility (Proximity to primary and secondary services) a major criterion for the planning of land use in New Tehri [17, 18].

Furthermore, the town has steep slope protection and sunlight exposure because the town has east, south, and south-east facing slopes, with multiple functions situated in layers and causing minimal disruption to the fragile hill ecosystem [19]. The planning of New Tehri is done in a way to increase the viability and sustainability of the town. However, the town has to face various challenges such as unchecked and unplanned development due to poor enforcement of the building bye-laws and inappropriate provision of the land uses. The newly built houses are away from the central area, having inadequate open spaces and less exposure to sunlight. Commercial buildings like shops are directly open to the roads and the built environment is dominating the natural environment of the town [20]. New Tehri also lacks tourism in spite of having good climatic conditions and topographical advantage due to a dearth of high-quality recreational facilities, shops, tourist accommodations, community facilities, and water scarcity that could entice visitors [18].

With the help of the above case study of three hill towns, namely Bilaspur, Lavasa, and New Tehri, a comparative analysis is done to understand the land use planning process of these hill towns which are discussed below (Table 1).

The three towns namely Bilaspur, Lavasa and New Tehri are examples of the hill towns of India established on new contour sites for rehabilitation due to the construction of dams. The case studies aided in understanding the issues associated with urban expansion in fragile hill areas. These hill towns are planned, designed and developed to cater for a fixed population. However, the strain of population expansion and development is always there in these places. The main problem that all three communities continue to have is improper land-use allocation without considering the site's context. In the case of Bilaspur town, data was not available for land use

Table 1 Comparative analysis of Bilaspur, New Tehri and Lavasa City

City	Bilaspur, Himachal Pradesh	Lavasa, Maharashtra	New Tehri, Uttarakhand
Year of establishment	Planning was done in 1960's in which the sectors were divided inside Nagar Parishad area and was a part of rehabilitation project	Planning of the town was done in year 1999, as a part of rehabilitation project due to construction of dam	New Tehri, was established to as a part of rehabilitation project and development was completed in 1993
Planning authority	The sectoral plans have been prepared by TCPO, Delhi	Lavasa Corporation Ltd. Planned for the development of a new hill town	The planning has been done by Government of Uttarakhand
Major achievement	Bilaspur is known as the first planned town of the Himachal Pradesh	The first planned hill city in Maharashtra	The Prime Minister's Award for Excellence in Urban Planning and Design
Area	Nagar Parishad area is 388 hectares	Development of new hill town area is 10,000 hectares	The town covers an area of 3705 hectares
Planning concept	The development was done in the sectors. (Sector planning)	Based on principle "transect model"	Developed on egalitarian ideologies (Improving the urban structure)
Criteria for land use planning	Factors that were considered for Land Use planning are: Available Developable land in each sector, Surrounding land uses, Threshold population, Locational attributes of amenities, Site characteristics, Work place relationship and Land values	Information for preparation of land use maps: Various surveys such as Ground survey, ecological survey, contour survey, terrain and Hydrological Information	There were five main criteria considered by the government for planning of New Tehri town: Functionality, sustainability of environment, urban design with cost effectiveness, Social and democratic factors

Source Compiled by author

planning, therefore a semi-structured interview survey with various town planning officials and people who had lived there for a long period of time was conducted to understand how the town developed. The land use allocation process for Bilaspur Town and New Tehri Town was not carried out using appropriate scientific methods or models. Instead, it was based solely on population projections for the future, and the professionals neglected to take into account the interdependencies between the adjacent land uses. This led to the haphazard development of these towns.

5 Study of Existing Land Use Planning at International Hills

Land use planning has become an essential tool worldwide for the development and management of the urban growth of nations. In order to accommodate changing needs, land use planning approaches must be flexible and responsive. The "blueprint approaches" that outline the phases, procedures, and tools do not work; rather, land use planning must be designed as a process that is suited to the site's specific needs, requirements, abilities, norms, and institutional structures [21]. Because of the difficult terrain and topography of the hills, planning must be more systematic. To understand how land uses are planned around the world, case studies of hill areas in Bhutan, Nepal, and the United States are taken and the various steps taken by these countries for land use planning are examined in the study.

5.1 Bhutan

Bhutan has an area of 38,394 km^2 and is located in the Eastern Himalayas. The majority of the population is subsistence farmers who rely on agriculture and livestock for a living [22]. In Bhutan, several different plans and legal frameworks have been developed to govern the land's usage in different contexts, such as in urban areas, agricultural settings, forested areas, and conservation areas [23]

Land use planning under NLUP in Bhutan: The aim is not to focus on a particular use of the land but to adopt multiple uses and overlap different categories on the land use map. The policy for coordination between these multiple uses that overlap each other is an important component of land use allocation in the NLUP [23].

The mapping is based on both spatial (LULC maps) and non-spatial (non-spatial) data collected from the concerned organizations. Steps taken for land use planning in Bhutan is shown in Fig. 3.

Thimphu, Bhutan: A structure plan was created for Thimphu, which included collaboration with stakeholders to ensure their ownership of the plan. Although no blueprints were created, the plan included management instruments for the expansion

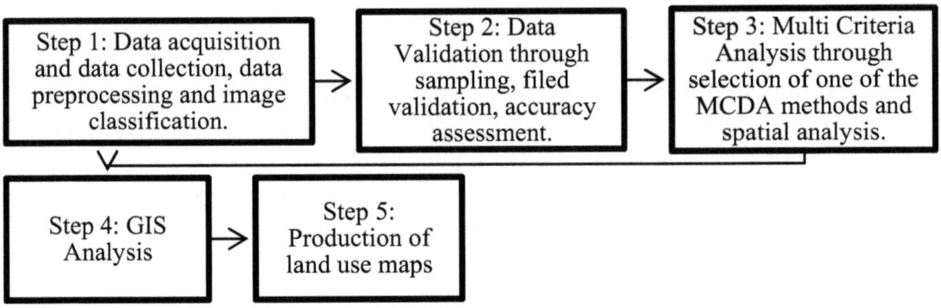

Fig. 3 NLUZ general technical flowchart. *Source* Compiled by authors and extracted from [24]

of Thimphu [25] The Principle of Intelligent Urbanism, which governs urban design, growth, and management, is the basis of the capital city [26]. Strategies adopted for land use planning in Thimphu, Bhutan are regionalized growth, Densification, transport-oriented growth, Resource utilization, Urban precincts, etc. [26].

5.2 *Nepal*

Nepal has 147,181-sq. km. of territory with a population of 23.4 million. Nepal has a diversity of plants, wildlife, and hilly as well as mountainous areas [27]. The land use categories in Nepal have been prepared by the National Land Use Project. A top-down strategy has been implemented [28].

Land use planning in Nepal: In Nepal, the land is allocated for various purposes based on the mentioned steps in Fig. 4.

Kathmandu, Nepal: Kathmandu is in the middle of Nepal's east side and occupies about 68,500 hectares of land. The district is composed of flat plains at 1300 m above mean sea level and it surrounds the mountains which range up to 2800 m tall. For land use planning, the strategies applied in Kathmandu, Nepal is:

- Planning at two Levels: Macro Level (Valley) & Micro (Municipal Level)
- Two-tiered approach to planning (Valley) and more detailed, micro-level (Municipal Level)
- Zoning based on constraints and risk sensitivity: In the land use plan, Multi-Hazard Vulnerability Analysis was used to look at the risk and decide how each piece of land should be used [30].

5.3 *United States*

According to the land use resource guide, the overall suitability index for the United States has been calculated using three criteria: unsuitable for development,

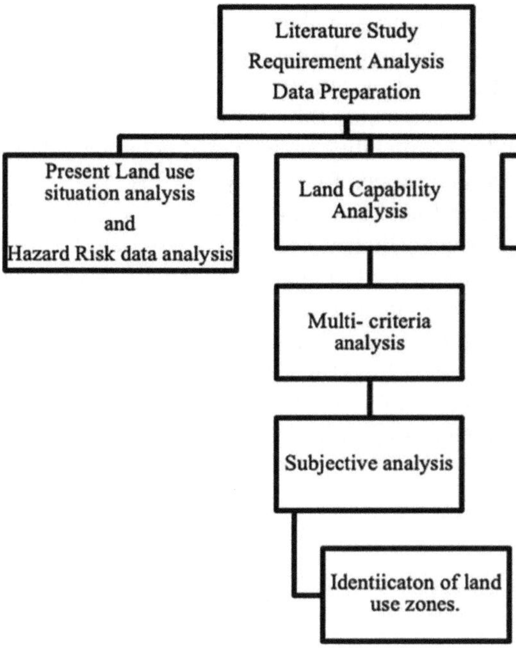

Fig. 4 Strategies for land use planning in Nepal. *Source* Compiled by authors and extracted from [29]

marginally suitable, and suitable. Land use planning in the United States is governed by a comprehensive plan in which land use elements contribute to a more transparent planning process.

Land use planning in US: Steps in preparing land use elements are described as follows included land use inventory, i.e., Preparation of base map by collecting the land use data, Land use Analysis, Land use goals and objectives by involving the public and framing out the goals, objectives and policies, Setting alternative land use scenario by understanding the alternative options for the future and checking its advantages and disadvantages, Preparing the future land use map: From public participation and above mention steps, prepare the future land use map, its implementation, monitoring and future updates [31].

Santa Clarita, US: Santa Clarita is a city in Los Angeles County, in the southern California region of the United States. Santa Clarita Valley is a village-populated valley.

For the land use allocation, firstly the comprehensive assessment and distribution of the existing land use are done through aerial photography, Survey of the field and GIS application. A land use suitability study is undertaken in which the various factors for city land use distribution are considered such as terrain, accessibility, closeness to infrastructure, environmental restrictions, the character of surrounding growth, economic feasibility etc. [32, 33]. Land use overlays are added to the base land use on the existing Land Use Map and best-suited use is designated to a particular piece of land. This effective tool is used in Telluride, Scottsdale, Pasadena, Los Gastos etc. [34]. Once the suitability analysis is performed, inputs from the public i.e., The public engagement process are conducted through seminars, email, and the city's website. Thus, through land use suitability analysis and comments through public participation a land use map is developed for the city [32, 33].

The literature on international land use planning summaries the processes used for land use in hill cities of Nepal, Bhutan, and the United States. It also provides details on the methods and surveys used to identify land development. These cities have a record of land use planning documents, and systematic land use planning is carried out in compliance with the land use documents established by the nations. The land use is assigned using any of the MCDA techniques with some predefined criteria based on the requirements of that hill city. The emphasis is also on land use suitability, which makes the process more effective in these hill cities of Nepal, Bhutan, and the USA.

6 Discussion

The comparison of national and international land use planning practices is based on the following considerations, which are described in brief below.

Cities:

In the case of India, the hill cities of Bilaspur in Himachal Pradesh, Lavasa in Maharashtra, and New Tehri in Uttarakhand were studied as they are considered to be India's planned hill towns. For the international hill cases, the hill cities Thimphu (Bhutan), Kathmandu (Nepal), and Santa Clarita, Anchorage, and Boone County (United States) are analyzed in order to comprehend the land use planning processes followed in these hill cities.

Guidelines:

URDPFI guidelines are followed in India for land use planning. Hill regions of India lack guidelines specific to hill development and there is a requirement for hill-specific provisions in these fragile areas. Whereas, in the case of Bhutan, Nepal, and United States hills, there are specific documents that can be employed as a reference for land use planning such as NLUZ (National Land Use Zoning, Guidelines), NLUP (National Land Use Project), and Land Use Resource guide respectively. In addition to this, they also consider the guidelines that are recommended by the FAO (Food and Agriculture Organization) for effective land use planning.

Documentation:

The documentation of the land use process in India is shown in the development plans of the various towns through the growth trend (GIS maps). The documentation of the process must be clear, including the basis on which the land uses are allocated, the technique used for land use planning, such as statistical evidence or any other analysis, the criteria considered etc. to guide to future planned development of these hill towns. At the international level, there is a brief documentation of the land use planning process. The documents provide descriptions and details of the procedures or methods used for land use planning, which aids in a better understanding of the planning process and serves to direct future development.

Data availability:

There is a lack of data for land use such as socioeconomic data, land use trend data, occupation distribution data, disaster risk data, multi-hazard data, spatial data, lack of local experts/professionals, funding and resources. Since the land use planning process was properly documented in the international cases, it was simple to find data in documents at the international level. The documents contain information on land use trends, process descriptions, etc. making the planning process more transparent.

Criteria:

Various considered for Land Use planning such as Available Developable land in each sector, Surrounding land uses, Threshold population, Locational attributes of amenities, Site characteristics, Work place relationship and Land values in case of Bilaspur, Functionality, sustainability of the environment, urban design with cost-effectiveness, Social and democratic criteria in case of New Tehri. However, in the case of Lavasa, Maharashtra, the criteria were not specified. The documents only mentioned the surveys used for the planning and development of the town. In the case of Bhutan, the criteria are institutional infrastructure (education, health, and other institutions), tourism, mining, heritage sites, heritage forests, and national parks, divided under the themes of social, economic, culture and environment. Nepal considers wetlands, abandoned land, sloping land, watersheds, high mountains, flood-prone river banks, barren lands, wetlands, abandoned land, unused or underutilized lands, If possible, along the sides of roads, canals, etc. To create natural pollution protection around industrial areas, Other appropriate locations for agro-forestry or timber products, etc., on the land with a high or medium risk of hazards. In the case of Santa Clarita, USA, the following aspects are taken into account: topography, accessibility, proximity to infrastructure, environmental constraints, the type of surrounding development, economic feasibility, etc.

Technique:

In case of Bilaspur the planning is done in sectors, the transact model approach is followed in Lavasa. The planning of New Tehri is done in a way to increase the viability and walkability in the town through construction in the terraces (levels). After selecting the criteria for the hill towns, the multi-criteria analysis and situational analysis are carried out as a scientific method to allocate the land uses at the international level that makes the planning in these towns efficient.

A comparison of land use practices is drawn from the literature of various hill cities of India and abroad. Table 2 compares the land use planning processes followed in the hill cities of India, Bhutan, Nepal, and the United States where * Means the

Table 2 The table shows the comparison of the land use planning process between India, Bhutan, Nepal and United States

Country	India			Bhutan	Nepal	United States
City	Bilaspur (HP)	Lavasa (Maharashtra)	New Tehri (Uttarakhand)	Thimphu	Kathmandu	Santa Clarita
Guidelines	#	#	#	*	*	*
Documentation	#	#	#	*	*	*
Data availability	#	#	#	*	*	*
Criteria	*	0	*	*	*	*
Technique	0	0	0	*	*	*
Tool	*	*	*	*	*	*
Referred document	0	0	0	*	*	*

* Means the characteristic is present or followed for the city
0 Means the characteristic is not present or followed for the city
\# Means the characteristic is present for the city but the information in the documents is incomplete

characteristic is present or followed for the city whereas 0 represents the characteristic is not present or followed for the city and # Means the characteristic is present for the city but the information in the documents is incomplete.

It is clear from the table that India lacks a systematic and organized approach to land use planning. These examples contribute to a better understanding of the importance of land use planning and management. Land-related issues in various parts of the country especially in the case of hill towns or cities point to shortcomings in planning. If these trends continue, India will fall short of the equity and resources that are present in these fragile areas. It is essential to understand that land serves numerous natural and human functions, and it is critical that it be used optimally and planning of the land uses must be carried out in a scientific and systematic manner [5].

7 Conclusion

The study examines spatial planning and land use on a national and worldwide scale. Due to the scarcity of land parcels in the hills and the rapid rate of development, the planning of the mountainous, sensitive areas is the main focus. The lack of knowledge and study on land use planning in hilly areas of India is a major barrier to attaining the goals of Sustainable Development Goal 11. The study of three planned Indian hill cities—Bilaspur, Lavasa, and New Tehri describes the concepts employed in the development of these cities, but its documentation does not provide the mechanism or method of allocation. Land use planning in the Indian hills is ineffective due to a paucity of information, a lack of data, improper documentation, the absence of a scientific basis for the allocation of land uses and the lack of a standard method for allocating land uses. These issues are hindering the progress of Hill Cities toward resilience and sustainability. In comparison, the hill cities of Thimphu (Bhutan), Kathmandu (Nepal), and Santa Clarita, Anchorage, and Boone County (United States) demonstrate that there is a systematic procedure followed along with proper documentation that leads to the planned development of these places, and thus making planning in these hill cities more efficient. This comparative analysis can contribute to the sharing of best practices, which is a key component of SDG 11.

Due to improper land use and a lack of land use planning, the hills of India confront a variety of issues and challenges making the cities less resilient. The lack of documentation regarding the land use processes for the hills and the absence of plans containing mechanisms for distributing land uses are worrisome. For the allocation of land use, there is no particular procedure that can be replicated in the hills. The records and documentation pertaining to the planning of land use in the hills do not include any criteria or methodology, making it difficult to tell how the optimal use is chosen or decided. In the Indian hills, urban development or the growth trend influences the land use allocation process, although it should be the other way around, i.e., the land use allocation process should dictate the growth trend of a particular region, which planners must comprehend. As a result, there is a need to design a method or procedure that may serve as a reference point for land use allocation in the hills of India, as the issue of land use planning is more pressing in hill towns of India, experiencing fast urbanization and unplanned development.

References

1. Amin A, Fazal S (2017) Assessment of forest fragmentation in district shopian using multi-temporal land cover (A GIS approach). J Geosci Geomatics 5(1):12–23
2. Bandyopadhyay R (1993) Land system in India: a historical review. Econ Polit Wkly 28(52):A149-A155
3. Johnson-Roehr SN (2014) City planning in India. Springer Science, pp 1–12
4. Airey J, Doughty C (2020) Rethinking the planning system for the 21st century. Policy Exch
5. Aijaz R, Knopf F (2019) Regional planning for sustainable land use in India. Obs Res Found Glob Policy J
6. Sharma S, Saini A, Shrivastava B, Kumar G, Kumar A (2022) Evolution of Indian hill stations during the British era: problems and prospects of development. ISVS e-J 9(2):114–129
7. Wang H, Shen Q, Tang B (2014) A review of planning support systems for urban land use planning
8. Van Herwijnen M, Janssen R, Stewart T, Aerts J (2005) Evaluating spatial design techniques for solving land-use allocation problems. J Environ Plann Manag, Taylor & Francis 48(1):121–142
9. Chawhan V, Kamal MA (2021) A study of planning, design and construction of buildings in hilly regions of India. Am J Civ Eng Archit:13–22
10. Department of Town and Country Planning. Development plan for various planning and special areas of Himachal Pradesh. Department of Town and Country Planning, Government of Himachal Pradesh

11. Sharma AS, Sood AM (2020) Analyzing the role of land use planning and development regulations in Solan city. J Emerg Technol Innov Res:104–112
12. Vishwapremi B (2016) Bhakra Oustees demand regularisation of titles and allocation of plots in Bilaspur, Himachal Pradesh. https://www.landconflictwatch.org/conflicts/regularisation-of-titles-and-allocation-of-plots-to-bhakra-dam-displaced
13. Mishra A (2019) Issues of large-scale dam resettlement and rehabilitation: case of Bilaspur, Himachal Pradesh. Indian J Public Admin:1–21
14. CEO Water Mandate (2022) UN global compact. Pacific Institute. https://ceowatermandate.org/
15. Hambarde A (2009) A GIS based master plan for hilly Terrain—Lavasa case study. Lavasa Corporation Ltd., Pune
16. Dutta AP, Shrivastava KS (2011) Lavasa exposed. https://www.downtoearth.org.in/coverage/lavasa-exposed-33282
17. Kaur H, Garg P (2019) Measuring sustainability of development in planned hill towns: a case of new Tehri town. In: Research into design for a connected world. Smart Innovation, Systems and Technologies, vol 134, Singapore, Springer
18. Kaur H, Garg P (2019) City profile: New Tehri. Cities
19. Kaur H, Garg P (2021) Case-based assessment of planned hill town using existing urban sustainability assessment tools. Environ Dev Sustain
20. Garg P, Kaur H (2019) Sustainability issues in context of Indian hill towns. Sustain Energy Build
21. Metternicht G (2017) Land use planning:1–66
22. Tshewang U, Tobias MC, Morrison JG (2021) Bhutan: conservation and environmental protection in the Himalayas. Springer
23. Ministry of Works and Human Settlement (2019) The project for formulation of comprehensive development plan for Bhutan 2030. RECS International Inc.
24. National Land Commission Secretariat, Royal Government of Bhutan (2018) National land use zoning, implementation guidelines 2018. National Land Commission, Kawang Jangsa, Thimphu
25. Ministry of Works and Human Settlement Royal Government of Bhutan (2008) Thimphu city development strategy. Ministry of Works and Human Settlement Royal Government of Bhutan, Thimpu
26. Benninger CC (2002) Principles of intelligent urbanism: the case of the new capital plan for Bhutan. Ekistics:60–80
27. Oli PP (2001) Spatial data for land use planning in Nepal. In: International conference on spatial information for sustainable development Nairobi, Kenya, Kenya
28. Nepal P, Khanal NR, Zhang Y, Paudel B, Liu L (2020) Land use policies in Nepal: an overview. Land Degrad Dev:2203–2212
29. Government of Nepal Ministry of Land Reform and Management. National land use project
30. KVD Authority. 20 Years strategic development master plan for KV (2015–2035). Kathmandu Valley Development Authority, Kathmandu
31. Land Use Resource Guide Team (2005) Land use resource guide. Center for Land Use Education, University of Wisconsin-Stevens point/extension
32. Santa Clarita Municipal Code (2011) Land use element, city of Santa clarita general plan. https://www.codepublishing.com/CA/SantaClarita/#!/SantaClaritaGP/SantaClaritaGP.html
33. Impact Sciences, Inc (2011) Final program EIR for the city of Santa Clarita's proposed one valley one vision general plan. Santa Clarita
34. Asheville Planning Department (2005) Hillside and ridgetop regulations: a recommendation for Asheville

Sustainable Land-Use Planning in Environmental Sensitive Hill City Gangtok, India

Sarika Bahadure, Tanushri Mokde Kamble, and Madan Lal Suthar

1 Introduction

1.1 Sustainable Land Use Planning

The fast pace of urbanization has led to a paradigm and devastating shift in land consumption patterns. Land use encompasses land for settlements, industrial sites, roads, infrastructure, and other human activities along with agricultural and forestry purposes [1]. Land use is one of the important sectors in urban development and has impacts on other sectors, and hence efficient land use planning is the prime need [2]. Sustainability in land use means preserving biodiversity and natural eco-balance while spatially distributing different uses. The identification and comparison of present and potential land uses for different land types, and the formulation and implementation of development programs based on these decisions provide opportunities for a sustainable space planning process [1]. Land is a limited natural resource so there is a need to assess the existing land use pattern and reorient future development to make it live able and sustainable.

21% of India constitutes of hilly or mountainous region [3]. Planning policies in India focus majorly upon the megacities [4] and thus smaller settlements and hill towns get lesser attention [5]. Most of the hill town are located in the ecologically sensitive zones [6]. They face issues like high-density, unplanned growth, congestion,

S. Bahadure
Department of Architecture and Planning, Visvesvaraya National Institute of Technology, Nagpur, Maharashtra, India
e-mail: sarikabahadure@arc.vnit.ac.in

T. M. Kamble (✉)
Institute of Architecture and Planning, Nirma University, Ahmedabad, Gujarat, India
e-mail: tanushri.mokde@nirmauni.ac.in

M. L. Suthar
Town Planning, Local Self Government, Government of Rajasthan, Jaipur, India

V. Chettry et al. (eds.), *Geoinformatics for Hilly Regions*, Advances in 21st Century Human Settlements, https://doi.org/10.1007/978-981-95-0156-4_4

urban sprawl, unauthorized development, regular weathering due to rainfall thereby affecting the sustainability [7–10]. They are rich in resources like water, petroleum, minerals, forests with rare flora and fauna [11]. However, due to the altitude and steep slopes, agricultural activities remain restricted [12]. With the rise in population, growth and development remains inevitable, and hilly areas are no exception. This results in alteration of topography, affecting the natural drainage and vegetation cover. The unplanned alterations in the fragile ecosystem can be fatal.

1.2 Planning in Hilly Areas

The hill area is a region with a height of more than 600 m above sea level or with an average slope of 30 degrees or more [13]. A large amount of attention has been paid to hill areas in recent years due to the planning challenges and the continuous demand for infrastructure in these areas [14]. The steep slopes and difficult terrain impose infrastructure provision, mobility, and connectivity concerns. Since terrain plays a crucial role in hilly areas, settlements must be established on definite habitable lands. For adequate planning, their carrying capacities must be determined. Currently, the planning status of most of the hilly areas is uncontrolled creating haphazard growth generated by urbanization, industrialization, and intrusion of commercial activities on the limited land cover mostly along hill routes. These trends have resulted in encroached forest land and shattered green cover. The construction activities and settlement expansion on unsuitable lands like the top of ridges, and steep slopes are disrupting the natural ecosystem and making these areas more vulnerable. This haphazard growth has impacted the infrastructure and resulted in traffic chaos, inefficient physical and social infrastructure, congestion in prime locations, and decay of natural resources. These have directed the environmental impacts, including loss of hills and forests, degradation of stream systems, landslides, erosion, and an increase in natural hazards including earthquakes, landslides, air pollution, and roadway noise [15].

Hilly areas should be dealt with special consideration as they have a sensitive and fragile ecosystem, which needs to be conserved [16]. It is essential to plan and develop strategies for hilly areas with additional sensitivity and a strong understanding of land use and settlement patterns. Thus, this paper studies the hill city Gangtok, which is a fast-growing, most populous, capital and the largest city of Sikkim state in India. Being a part of Himalayas, it is rich in biodiversity. The state has around 35% of land under Khangchendzonga National Park and houses nearly half of Nations biodiversity [17] and contains more than 100 glaciers [18]. The phenomenon of urbanization in Sikkim is Gangtok centric which reinforces the imbalanced urban structure & primacy of Gangtok city.

The city has undergone tremendous changes in the last three decades imposing challenges toward sustainability. Over the last three decades, the increasing population, urbanization, and increasing tourist activities have imposed issues like scarcity of buildable land, susceptibility to major mishaps, poor infrastructure, and deteriorating environmental quality which needs sustainable initiative. So, this study aims to assess the existing land use land cover distribution, planning process, characteristics of wards and city, creates a suitability map guiding the future urban growth with respect to the carrying capacity, and present recommendations for sustainable planning.

2 Methodology

The literature study is done to understand the urban development patterns, prescribed guidelines, and standards in environmentally sensitive hill towns. Gangtok city data is collected from both primary and secondary sources at ward and city levels. A reconnaissance survey was done to examine the general character of the city as per the ward characteristics. Primary data is gathered through household surveys (questionnaires by collecting 25–30 Samples from each ward), transportation survey interviews with local urban bodies, and surveys to update the existing land use plan. The secondary data sources were research papers, reports, administrative data, standards, and census data. It identifies different planning interventions in the form of development plans, land use plan/s, building regulations [16], suitability, and vulnerability analysis. The data processing involved creating maps like land use land cover, landslide vulnerability, slope map, etc. in the Geographic information system (GIS) from the satellite imagery and surveys, updating the existing land use, and computing the carrying capacity of the city. The analysis of various sectors like land use, demography, infrastructure, accessibility, and transportation was done to identify the issues, potentials, and thrust areas. The suitability map was generated based on physical and social criteria further divided under 11 indicators. This helped to study the areas where development is needed and which areas need special management and protection. In the end, the recommendations are made for future planning and the study is concluded (Fig. 1).

3 Study Area, Gangtok, Sikkim, India

Gangtok is a hill city with a population of around 1,00,286 persons (census 2011) accommodated in 19.2 square kilometers of area. It has the scope of development, as the capital and the largest city of the Indian state of Sikkim. The city is located at an altitude of 1,800 m (5,800 feet) and lies at 27.33°N 88.61°E. The city is fringed on the east and west by two rivers, namely Rorochu (with 41 streams) and Ranikhola (with 44 streams) which divide the natural drainage into two parts.

Gangtok's existence is traced back to the old silk route in the first century. In the eighteenth century, Gangtok gained importance as a Buddhist Pilgrimage town and became a Capital in 1894. The monarchy of Sikkim was withdrawn in 1975 and it

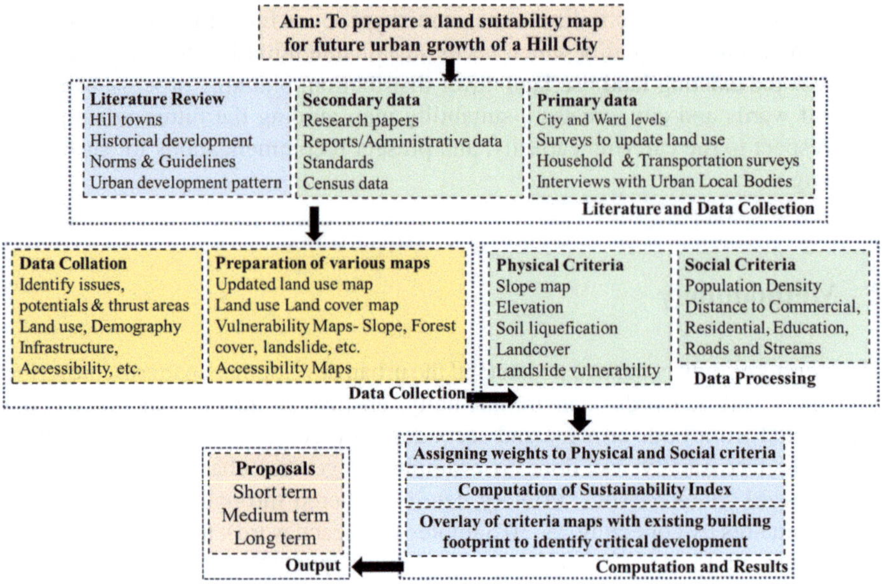

Fig. 1 Methodology chart

became an Indian state (see Fig. 2). Since then, the road up-gradation (formation of National and State Highways) and the development of multiple tourist locations and infrastructure increased the footfall of the tourist.

Gangtok is not just the largest city but also the primate city of Sikkim as more than 65 percent of the total urban population stay in Sikkim where the factor of primacy is as high as eight [19, 20]. Gangtok has consistently recorded a very high population growth rate which is more than the national growth rate except for the decade 1981–91 which recorded a negative growth rate of 31.90 percent [19]. The decade 2001 to 2011 has shown the most phenomenal growth rate of 129.43%. This huge increase in the Gangtok population can be due to the reorganization of the Gangtok Municipal Corporation (GMC) boundary [21].

Gangtok's growth has been accelerated, due to the increased economic opportunities generated owing to tourism and the primate city [19]. The steep slopes, vulnerability to landslides, large forest cover, and inadequate access to the majority of areas have hindered the natural and balanced growth of the city [21]. There is a need to identify and open the underdeveloped areas within the notified urban area for development through better infrastructure, accessibility, and transport measures. The future direction of growth needs to be spread over a larger area taking into consideration the development constraints like topographical features (landform), watercourses, streams (jhoras), forests, and growth propensity which demand the assessment of city land suitability.

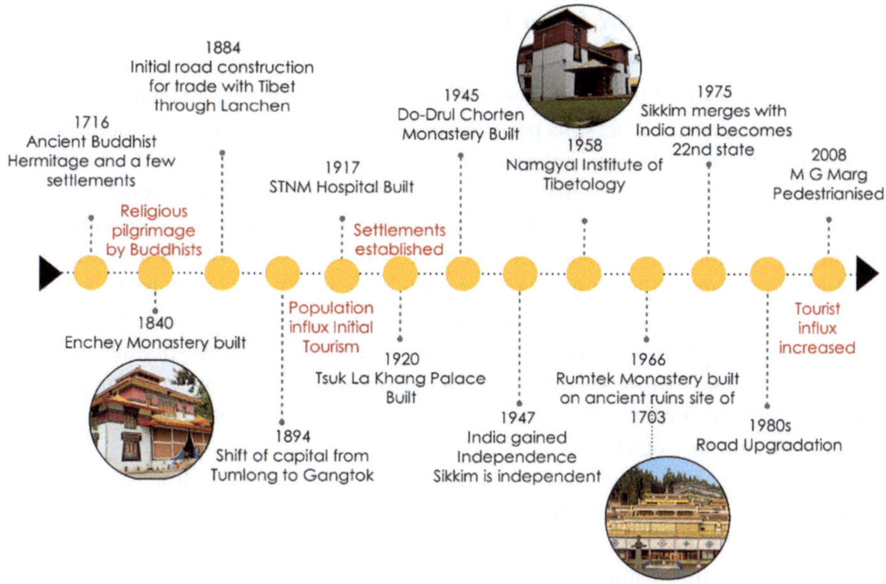

Fig. 2 History and evolution of Gangtok

3.1 Spatial Growth, and Land Cover

Gangtok, exhibits a linear pattern of development, along the major transportation route [20]. The primary growth axis is in the South and Southwest direction, along the NH 31A on the Ranipool axis as the terrain in this axis is relatively more favorable [21]. The disorganized land use changes seriously affect the biodiversity of local habitats. Even though there exist difficulties like the presence of slopes, the vulnerability of landslides, and inaccessibility, then also many areas have been diverted for habitation. Paul et al. [22]conducted land cover analysis from the satellite imagery for the years 2001 and 2011in GIS. It shows that pristine and eco-sensitive natural reserves are encroached upon for human habitation and allied activities for urban development. Mixed build area (urban) and build area (residential) increased from 12.54 to 14.85 and 29.86 to 38.02 percent respectively. However, the forest semi-evergreen and tree plant area reduced from 45.60 to 37.69 and 6.02 to 5.00 percent respectively from 2001 to 2011 [22]. Already, the city is located in an ecologically fragile and seismologically vulnerable zone [23]. This rapid land conversion and unplanned growth may pose risk to the people and society.

3.2 Ward Profile and Land Use

The Municipal limit of the city is divided into 15 wards (see Fig. 3a) with an average population density of 52 persons per hectare (pph), the highest and lowest density of 301 and 19 pph respectively. Figure 3b represents the population density distribution of the city. Northern wards have a sparse population featuring eco-sensitive zones and natural tourist destinations. These tourist destinations have led to an increase in commercial establishments such as hotels over the years. The Central wards along with the core feature dense commercial centers. The residential land use is seen only in the form of mixed land use (commercial + residential). They also have a high concentration of administrative buildings and other public and semi-public buildings. The Southern wards show a typically dense residential land use with commercial use seen in the mixed format. They feature institutional buildings such as hospitals and educational campuses as well.

Gangtok does not have ideally planned urban areas in their true sense, as the guiding factor for development has been accessibility and availability of suitable land profiles. This has resulted in a varied land use pattern with no allocated specific activities as per the available resources [24]. Industrial land use is not observed within the city limits, except for a few small and medium-scale industries in south at Tadong and Ranipool and metal fabricator units in Chandmari at the east. Recreational land use is a deficit with underprovided organized open spaces, parks, and gardens. Commercial and mixed land use is majorly observed along the National Highway and arterial roads, while the interior roads served the residential land uses. Public/semi-public land use is distributed throughout the city. From 1975 to 2011 residential sector decreased by 29 percent, whereas the transport, recreational, public/semi-public, and industrial sectors increased by 7%, 1.7%, 20%, and 4% respectively [22]. Figure 3b represents existing land use which was computed through the onsite city survey.

4 Land Suitability Analysis

Land suitability refers to the ability of land based on certain characteristics to support a specific use. Its goal is to identify the most appropriate spatial pattern for future land uses conferring on specific activities, requirements, parameters, etc. based on land use, land cover, topography, and other land features. In planning process, land suitability analysis assists to identify the most appropriate land for particular land use [25]. Land suitability is done by evaluation and grouping of particular land areas based on their suitability for a defined use [15].

For assessing land suitability, physical and social criteria (Table 1) are evaluated. The data are analyzed and maps are prepared for the five physical indicators (see Fig. 4g–k) (slope map, elevation, soil liquefaction, land cover, and landslide vulnerability). The maps were overlaid with the existing building footprint to identify the areas of critical development. The social criteria assessed are population density, and

a) 15 Wards

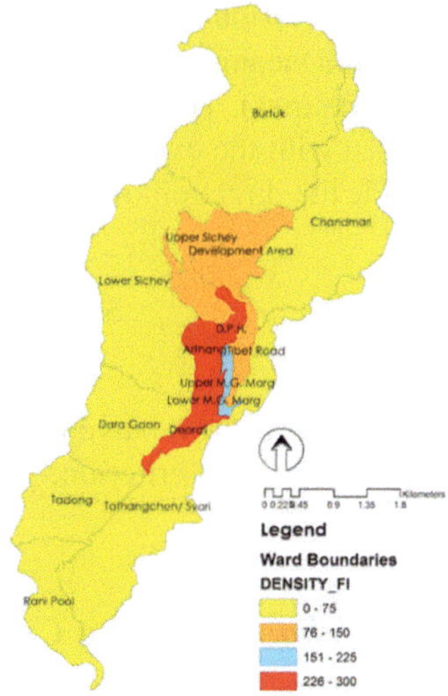

b) Ward Population Density Map

Land Use Category	Area in Sqm
Residential	671876
Mixed	240581
Commercial	346900
Industrial	1816
Public/Semi-Public	246265
Recreational	20917
Transportation	308887

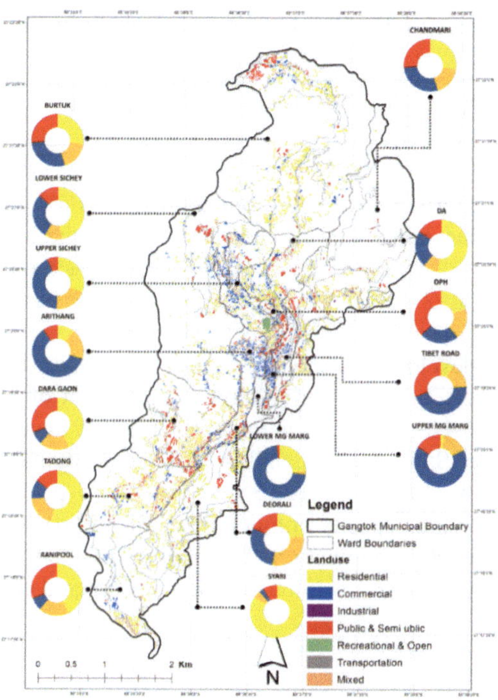

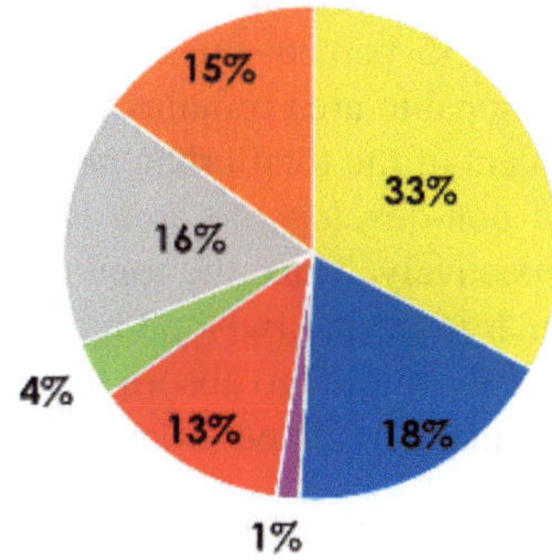

c) Land use distribution as per wards

Fig. 3 Gangtok-wards population density and land use maps distance to various destinations (see Fig. 4b–f) (commercial, educational, residential, roads, and steams). Each indicator is weighted and ranked (based on benchmarking and guidelines). The suitability index is computed with Eq. 1. The maps are superimposed with the existing built maps to create the land suitability map (see Fig. 4l) to guide the direction of future growth of the city.

$$\text{Suitability Index (SI)} = \sum_{i=0}^{n} (W_i * X_i) \tag{1}$$

where W_i is the relative weight of each criterion, and Xi is the standardized score/rank (0–3) of each criterion.

4.1 Urban Carrying Capacity-Gangtok

Carrying capacity refers to the maximum number of people that an area is capable of supporting with the resources available to it [26]. It can be calculated by Eq. 2

$$\text{CC} = \text{A}_{\text{U}}(\text{A}_{\text{ND}} + \text{A}_{\text{IF}}) \times \frac{\text{FAR}}{\text{S}} \tag{2}$$

where, A_U is the total delineated urban area (1934.8 ha), A_{IF} is the area under infrastructure development which includes areas under commercial, public semi-public, recreational, industrial, and transport sectors (97.29 ha), A_{ND} is the total non-developable area, i.e., 75 percent of A_U (1451.1 ha), FAR is floor space ratio (in hill towns which varies from 1.25 to 2.0) and S is the built-up area per person $\{(152.82 \times 1.5)/100286 = 0.002286 \text{ ha}\}$.

For the state of Sikkim, the non-developable area is 85, and the developable area is 15 percent out of which 12 percent is for agriculture, thus, percent of the developable area remains for urban development is 3 percent. Since more than 65 percent of the total urban population of Sikkim is in Gangtok [20], the developable and non-developable area for Gangtok is assumed as 25 percent and 75 percent respectively. Thus, the total urban carrying capacity computed is 2,53,550 persons which is around two and half times the 2011 population. This implies that the city has the potential to absorb the growing population without considerable degradation or damage to the environment.

Table 1 Criteria, indicators, weights and ranks

Criteria	Sr. no in Fig. 4	Indicators	Weights		Ranks (0–3)			
					3	2	1	0
Social	Figure 4a	Population Density	0.4	0.30	<60	60–90	>90	
	Figure 4b	Distance to Commercial		0.15	<400	400–800	>800	
	Figure 4c	Distance to Educational		0.12				
	Figure 4d	Distance to Residential		0.18	<300	300–800	>800	
	Figure 4e	Distance to Roads		0.13	<200	200–600	>600	
	Figure 4f	Distance to Streams		0.12	<500	500–100	>1000	
Physical	Figure 4g	Slope Map	0.4	0.4	0–30	30–45	45>	
	Figure 4h	Elevation		0.1	2000>	1400–2000	800–1400	800<
	Figure 4i	Soil Liquefaction		0.1	Low	Medium	High	Very high
	Figure 4j	Land cover		0.2	Developable		Developed	Non-developable
	Figure 4k	Landslide Vulnerability		0.2	Low	Medium	High	Very high

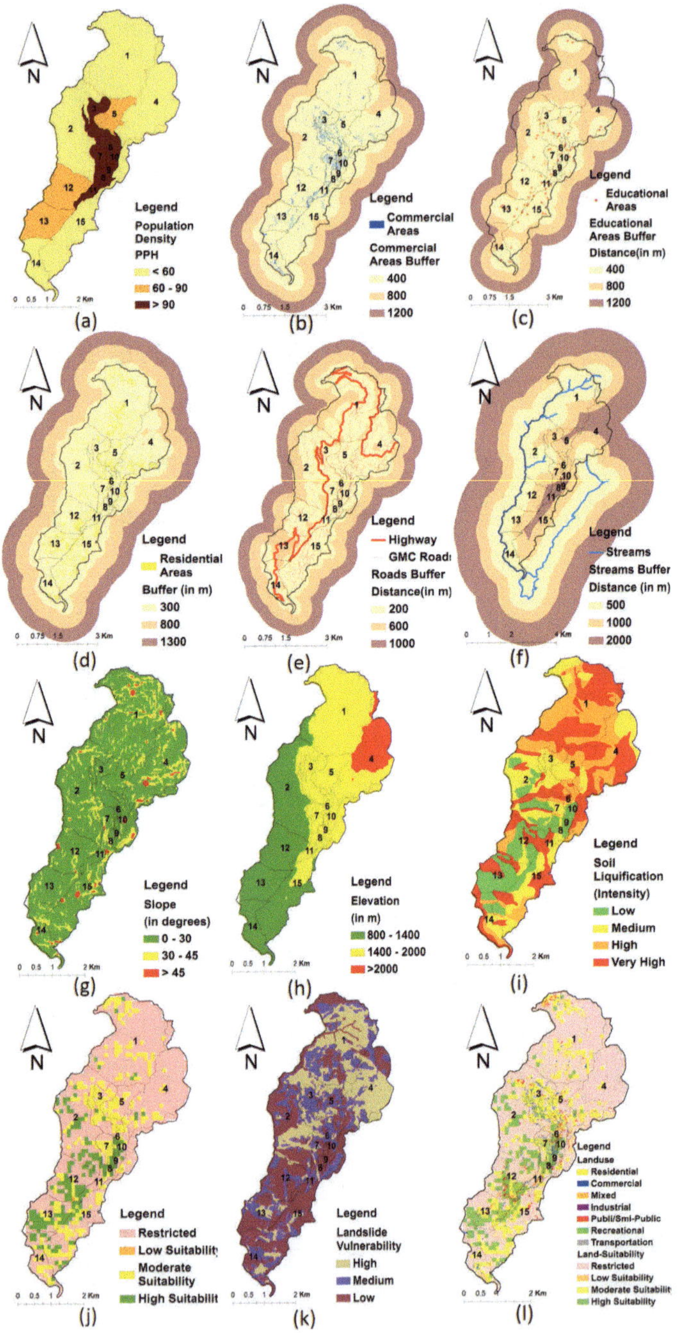

Fig. 4 Physical, Social Indicators maps and Land Suitability map

5 Recommendations

Short-Term Proposal—Considering the lack of recreational and open spaces in the city, there is a critical need to fulfill these gaps to cater to the health and well-being, and quality of life of the inhabitants of the city [27]. The requirement of the number of housing area parks and neighborhood parks can be estimated as per the URDPFI Guidelines.

Medium-Term Proposal—URDPFI standards suggest a density range of 60–90 pph in a medium-size hill town. Currently, CBD (Central Business District) LMG Marg, has the highest density of 282 pph, followed by Arithang and UMG Marg having densities of 235 pph and 223 pph, respectively. This has led to increased pressure on the infrastructure and the area is sustaining beyond its carrying capacity. Hence it is required to decongest the region by proposing a Sub-CBD for the city and decentralizing the activities. According to the land suitability mapping, the Tadong region toward south city is best suited for proposing a Sub-CBD, significantly away from the current CBD and with various opportunities for multi-disciplinary development, in the surrounding low-dense wards of Syari and Ranipool. Sub-CBD at Tadong would have a direct impact on the land use and infrastructure pattern of the Ranipool, Daragaon, and Syari region and an indirect impact on various sectors throughout the city. The activities in the core area would gradually be decentralized to Tadong and the adjacent areas. The Northern wards of Burtuk and Chandmari have low land suitability, higher elevation, low built density, and ecologically sensitive areas. Thus, it is proposed to limit the development in these wards and develop them as eco-sensitive areas, catering to the eco-tourism of Gangtok (Fig. 5).

Long-Term Proposal—From the literature study for Greater Gangtok, it is intended to develop growth centers in the peri-urban areas of Gangtok. Pakyong, Pangtang, Ranka, Luing, Bhusuk and Assam Lingzay are the strategically chosen peri-urban areas within a distance of 30–35 kms from the city and are well interconnected. Pakyong- with the air connectivity, this area caters to service-related activities. Pangtang, Ranka, and Luing could be developed along with transit systems similar to Gangtok. Bhusuk- region being a fragile ecosystem and unsuitable for high-density growth, this could be developed as a recreational zone. Assam Lingzay should ideally be planned as a residential area. Properly planned development of the city limits would reduce the negative impact on the surrounding region, environment, and resources, making the settlements sustainable for the long term.

6 Conclusion

Land planning in hill cities is a crucial task and it should be done considering the fragile ecosystem, physical, and social vulnerability aspects. This study explores the spatial planning for a hill city Gangtok and provides a framework for sustainable future planning through Land suitability analysis and computing carrying capacity.

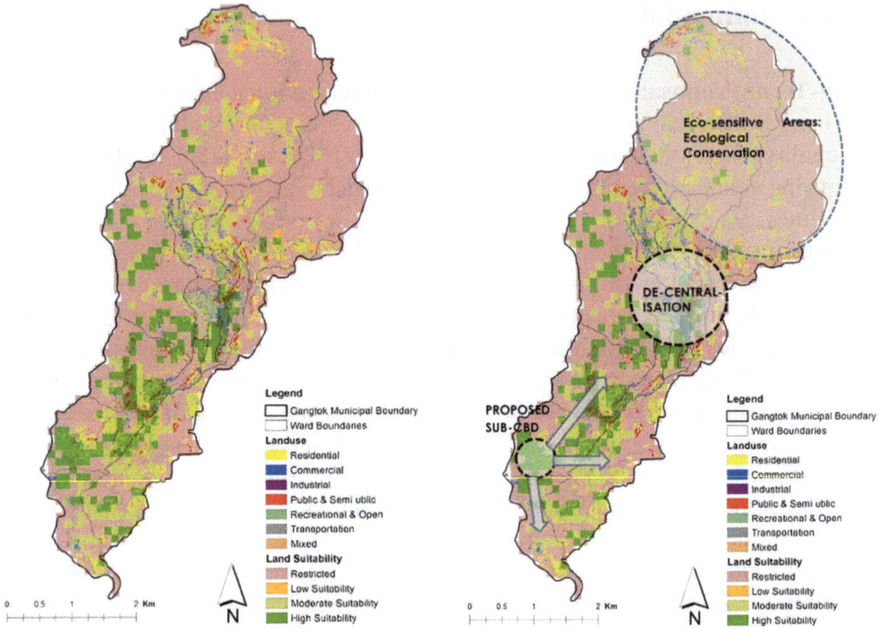

a) Proposed Open Spaces in the City

b) Proposed Sub-CBD and Decentralization of existing CBD

Fig. 5 Illustrative Recommendation (short and medium term)

The study generates a GIS-based suitability analysis considering land use land cover, physical, and social factors and provides recommendations for the small, medium, and long-term development. This framework fills the gap by providing a lucid qualitative and spatial planning process. This will assist the decision makers and planners to solve the issues faced by hill cities in a rational and effective manner. This study will pave the path for the other hill cities to take the decision for sustainable planning.

Acknowledgements The author thanks the students of the M. Tech. Urban Planning program at Visvesvaraya National Institute of Technology, Nagpur (India) for conducting onsite surveys, and compiling the data as a part of the Planning Studio in the summer 2019.

References

1. Wit PD, Verheye W (2009) Land use, land cover, and soil sciences. In: Land use planning for sustainable development, vol III. Encyclopedia of Life Support Systems (EOLSS)
2. Tadas G, Padhye A, Bahadure S (2021) Sustainable land use planning for Puducherry, India. In: International conference on resilient & liveable city planning (RLCP 2020), Vijaywada

3. Jain P, Singh I, Jain M (2021) Urbanization and land-use change in the hill town of Nainital. Mater Today Proc
4. Rajvanshi A (2003) Promoting public participation for integrating sustainability issues in environmental decision-making: the Indian experience. JEAPM 5(03):295–319. https://doi.org/10.1142/S1464333203001395
5. Mell IC, Sturzaker J (2014) Sustainable urban development in tightly constrained areas: a case study of Darjeeling, India. Int J Urban Sustain Dev 6(1):65–88. https://doi.org/10.1080/19463138.2014.883994
6. Kapoor M, Kohli K, Menon M (2009) India's notified ecologically sensitive areas (ESAs): the story so far... Delhi & WWF-India, Kalpavriksh, New Delhi
7. Sudhira HS, Ramachandra TV, Bala Subrahmanya MH (2007) Integrated spatial planning support systems for managing urban sprawl. In: Presentation at 10th international conference on computers in urban planning and urban management (CUPUM), Iguassu Falls
8. Shekhar (2011) Urban sprawl and other spatial planning issues in Shimla, Himachal Pradesh. Institute of Town Planners, India 8(3):53–66
9. She Y, Shen L, Jiao L, Zuo J, Tam VW, Yan H (2018) Constraints to achieve infrastructure sustainability for mountainous townships in China. Habitat Int 73:65–78. https://doi.org/10.1016/j.habitatint.2018.01.009
10. Kumar A (2016) Impact of building regulations on Indian hill towns. Hbrc Journal 12(3):316–326. https://doi.org/10.1016/j.hbrcj.2015.02.002
11. Misra JP, Mishra AK (2006) Sustainable development of agriculture in north eastern India: a quest for more economical and resources sustainable alternatives. Indian J Agric Econ 61(3):345
12. Groetzbach E (1988) High mountains as human habitat. In: Human impact on mountains. New Jersey, Rowman and Littlefield
13. NBC (2016) National building code of India 2016. Bureau of Indian Standards (BIS), GoI
14. Kapoor N, Jain M, Bansa V (2019) GIS-based framework for local spatial planning in hill areas. Int Rev Spat Plann Sustain Dev 7(4):113–130
15. GoI (2015) Urban and regional development plans formulation and implementation (URDPFI) guidelines, vol I. Government of India, Ministry of Urban Development
16. Kumar A, Pushplata (2015) City profile: Shimla. Cities 49:149–158
17. Sharma R (2019) Practices and challenges of municipal solid waste management in Gangtok town. Res Rev Int J Multi 4(4):1212–1214
18. Kumar P, Sharma MC, Saini R, Singh GK (2020) Climatic variability at Gangtok and Tadong weather observatories in Sikkim, India, during 1961–2017. Sci Rep 10(1):1–12. https://doi.org/10.1038/s41598-020-71163-y
19. GOS (2020) Formulation of GIS-based master plan for Gangtok 2041 (Under AMRUT scheme). Urban Development Department, Gangtok, Sikkim Govt. of Sikkim
20. AMRUT (2015) Sikkim: state annual action plan (SAAP) for FY 2015–16
21. Slum-Free City Plan of Action for Gangtok (2014) RAY cell: urban development & housing department. Government of Sikkim, Gangtok
22. Paul K, Sharma D, Mukherjee R, Sengupta R, Tamang KL (2016) Demographic characteristics and changing land use pattern in Gangtok, Sikkim (1975–2011). Int J Geomatics Geosci 6(4):1769–1780
23. Kaur H, Sarkar R, Gupta S, Parkash S, Thapa R, Meena SR (2022) The vulnerability of human population to landslide disaster: a case study of Sikkim Himalayas. In: Sarkar R, Shaw R, Pradhan B (eds) Impact of climate change, land use and land cover, and socio-economic dynamics on landslides. Springer, Singapore, pp 319–333
24. GoS Urban Development and Housing Department (2007) City development plan-Gangtok City. Gangtok
25. Puntsag G (2014) Land suitability analysis for urban and agricultural land using GIS: case study in Hvita to Hvita. Iceland
26. Sarma A et al (2012) Urban carrying capacity-concept and calclulation. Center of Excellence Integrated Landuse Planning and Water Resources Management (ILPWRM) Ministry of Urban Development, Govt. of India

27. Bahadure S, Sable S, Umbarje A, Vijayan M (2021) Public spaces and their place-making in Gangtok City, India. In: International conference proceedings on resilient & liveable city planning (RLCP 2020)

The Impact of Culture, Climate, Topography, and Geology on Settlements in Hilly Areas: A Case of Kandad, India

Sameer Gujar, Vaibhav Kulkarni, Amit Maurya, and Swati Bahale

1 Introduction

Urbanization exerts a great influence on settlements leading to altering their character and dynamics on a fundamental level. Although urbanization fosters economic development, innovation, and cultural diversity, it puts a massive strain on the existing infrastructure and increases social inequalities due to increased population [1]. Managing urbanization requires thoughtful planning, investment in infrastructure, and appropriate practices to ensure that these settlements can accommodate the changes while maintaining their unique character and promoting resident's well-being. Moreover, the United Nation has set a goal of enhancing inclusive and sustainable urbanization to promote integrated and sustainable, inclusive, safe, and resilient human settlements.

Population growth presents a complex set of challenges in hilly regions. These are characterized by their fragile ecosystems, tough topography, and high susceptibility to natural disasters like landslides and flash floods [2]. As families expand, the demand for housing, agriculture, and other natural resources increases, often leading to deforestation, soil erosion, and aggravating environmental degradation. In

S. Gujar (✉)
School of Architecture and Planning, NICMAR University, Pune, Maharashtras, India
e-mail: sameer.gujar@pune.nicmar.ac.in

V. Kulkarni
Padmabhushan Dr. Vasantdada Patil College of Architecture, Pune, Maharashtra, India
e-mail: ar.vaibhavkulkarni@gmail.com

A. Maurya
Department of Architecture, Planning and Design, IIT(BHU), Varanasi, Uttar Pradesh, India

S. Bahale
Department of Architecture, College of Engineering, Sungkyunkwan University, Suwon, Republic of Korea
e-mail: swatimbahale@gmail.com

V. Chettry et al. (eds.), *Geoinformatics for Hilly Regions*, Advances in 21st Century Human Settlements, https://doi.org/10.1007/978-981-95-0156-4_5

addition, these regions face difficulties concerning the provision of healthcare and education due to topography and remote location [3, 4]. Thus, balancing population growth with sustainability practices and preserving resource management are critical in hilly areas to retain their ecological integrity without compromising the delicate balance of nature [5, 6].

Settlements in hilly regions face multiple unique challenges stemming from their topography, geology, and climate. The rugged terrain of these areas can make construction and transportation difficult [7]. The unstable geology poses significant risks to settlements where managing these geological hazards requires careful planning, architectural, and engineering expertise. The climate in hilly regions can also be unpredictable, with variations in temperature and precipitation. Harsh winters, heavy rainfall, or prolonged droughts can impact agriculture and infrastructure, making it challenging for communities to sustain themselves [8, 9]. Moreover, the steep slopes in these areas can intensify the effects of extreme weather events, leading to soil erosion and flooding.

In essence, settlements in hilly regions must address the interaction of topographical, geological, and climate factors to ensure the safety, resilience, and sustainability of their communities. This often entails adopting innovative building techniques, implementing disaster preparedness measures, and promoting sustainable land use practices to mitigate inherent challenges.

The current situation of cities in hilly regions in India is complex and shaped by a variety of factors. Urbanization and rapid economic development have stimulated increased urban sprawl and infrastructure expansions in these areas [10, 11]. While this has increased the living standards, it also poses a significant challenge. The steep topography and unique geological characteristics make the settlements susceptible to landslides and other natural disasters. In addition, the effects of climate change, like unpredictable rainfall patterns and extreme weather conditions, further worsen these challenges. However, recent events indicate that the traditional settlements are rarely affected by these natural calamities. Thus, there is a pressing need for a detailed comprehensive study of existing settlements in hilly regions of India [12, 13]. Therefore, the current study identifies how these traditional settlements respond to the challenges of planning and construction in hilly regions.

2 Methodology

The focus of the current study is to understand how traditional systems have stood the test of time facing grave challenges faced due to the location and climate. The study analyzes an Indian hilly village, Kandad, situated in Uttarakhand state of India, in the lower Shivalik mountain range, an outer range of the Himalayas. Based on earlier studies, the topography and hydrological analysis is conducted in detail using geoinformatics at macro scale with the help of geographic information system (GIS) [14–16]. Topographic and hydrological mapping on ArcGIS platform involves using the SRTM dataset to generate slope, aspect, and triangulated irregular networks (TIN), a 3D digital representation, and delineate watersheds. Slope and aspect analyses provide critical insights into terrain steepness and orientation, influencing factors like erosion. TINs offer detailed terrain modelling, essential for various environmental and construction applications. The elevation values from Google Earth image helped develop the TIN model in ArcGIS. Similarly, land use development is tracked by conducting mapping of built areas over time.

Change detection in the built environment is performed by digitizing Google Earth raster images from 2011 and 2014 into vector layers, allowing for precise analysis of urban expansion and land use changes. Additionally, average annual rainfall mapping uses precipitation data from the Center for Hydrometeorology and Remote Sensing (CHRS) for the period of 2014–2021, essential for understanding hydrological cycles (Fig. 1). Since the study area is very small, there are limitations with acquiring related satellite data. Therefore, larger scale data was used for regional analysis and inferences are derived, respectively. Further, the study carries out a detailed micro analysis of construction practices and techniques and their response to local topography, climate, and culture by conducting documentation of structures in the study village. This is done through field surveys, photography, and measurements. It examines how these practices adapt to local topography, ensuring slope and erosion control, and how they respond to climatic conditions with weather resilience and natural ventilation techniques. The study also undertakes insights from focused group discussions and interviews with residents to understand the challenges they face and how they tackled them. This comprehensive approach provides a holistic understanding of how construction practices interact with the environment and cultural heritage, informing sustainable and culturally sensitive buildings.

3 Results and Analysis

3.1 Location and Geographical Analysis

Kandad village which is a part of the area known as Jaunsar-Bawar, is located on a ridge of a mountain range in the lower Shivaliks near Chakrata tehsil of Dehradun district in Uttarakhand, India. It is situated 16 km away from Chakrata and 109 km away from Dehradun and perched at a height of about 2,300 m above sea level (Fig. 2). With snow-covered peaks, glaciers, steep canyons, roaring streams, and a few patches of dusty plains, its terrain is extremely diverse. There are mountains nearby that are among the tallest in the world. As of 2011 India census, Kandad village has a total geographical area of 3 square km. There are 45 households and has a population of 265 of which males are 146 (55%) and females 119 (45%) [17]. The entire community stays in the village and is agrarian. Simultaneously, animal husbandry is one of the major sources of income. Cattle are used by them for their daily necessities. It includes caring, breeding, feeding, selling of the goats, cows, sheep extensively.

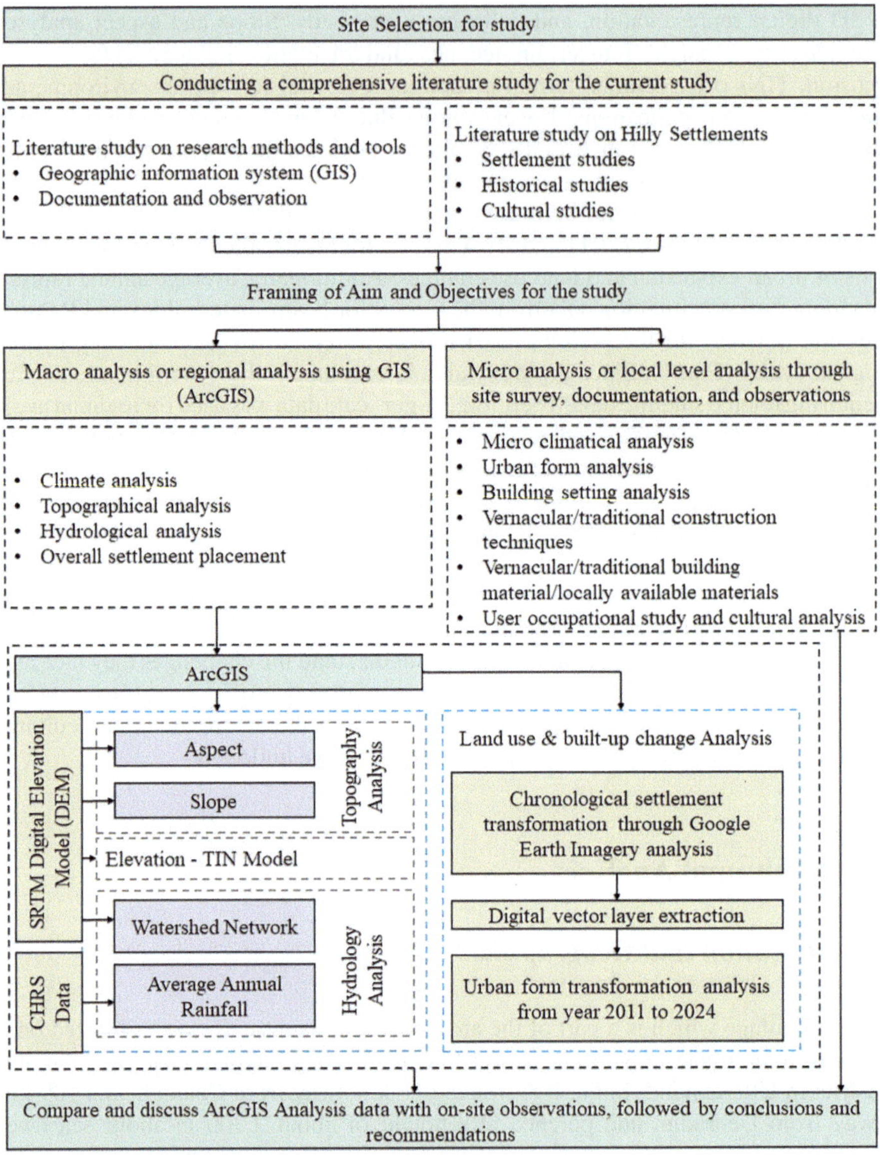

Fig. 1 Methodology for the study (*Source* Author)

The weather in Uttarakhand is influenced by a large extent by the geography. Mainly mountainous, the snowy peaks of Himalayas act as a natural cooler. The hills are known for their severely cold winters and almost non-existent summers. Kandad falls under the humid subtropical (Cwa) climate classification as per the Köppen-Geiger classification. The climate varies from tropical to severe cold, where

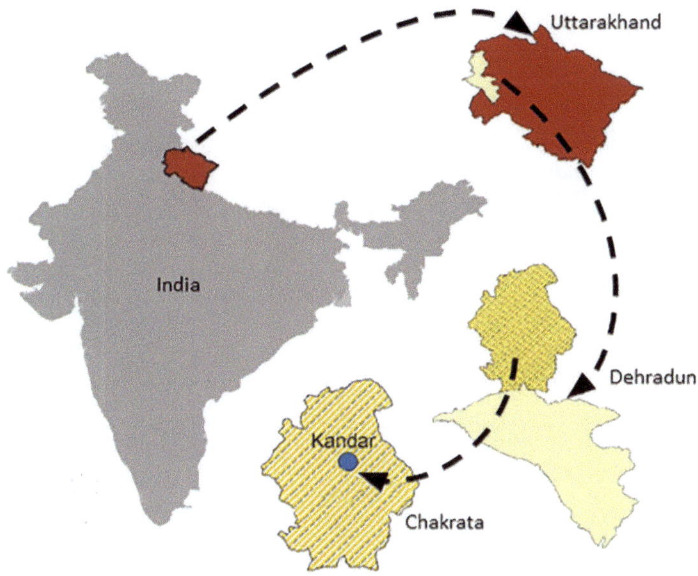

Fig. 2 Location of Kandad village (*Source* Author)

the cold months range from October (Minimum 10 °C) to March (Minimum 6 °C). The severe cold months range from December to February where the temperature falls to a low of −2 °C as it is located at a very high altitude. On the other hand, the summer temperatures go to a maximum of 30 °C. The region receives a medium rainfall with an average of 750mm precipitation amounting to snowfall (Fig. 3).

A watershed analysis indicates that Kandad village is located on the northern bank of a major water stream (Fig. 3), Since the precipitation is higher in the northern region of the village, the streams flood heavily during the monsoon season. The stream discharge is highest during the monsoon season and decreases gradually in the post-monsoon season. Similarly, several smaller streams pass through the village (Fig. 4). Moreover, the upper region has mountain soil, which is typically shallow and rocky, while the lower region has alluvial soil. Also, the slope analysis indicates a steep terrain which makes the area susceptible to soil erosion and landslides (Fig. 5).

Most of the houses of Kandad are more than 100 years old and have stood the test of time. The village settlement is built along the contours, where it spreads from south-east to north-west (Fig. 4). There are two sets of settlements, one located at the lower plains and the other located at the upper plains of mountains (Fig. 5). Each plain offers a unique advantage in terms of suitability for growing specific crops (at specific seasons on respective plains); rearing of specific cattle that can live on upper and lower plains and places of habitation for people. The entire effort ensures that their livelihood is not compromised despite the harsh conditions during the months of winter and the accumulation of snow. A typical family unit may own a couple of cattle and farmlands and places of habitation on both terrains. The same family will

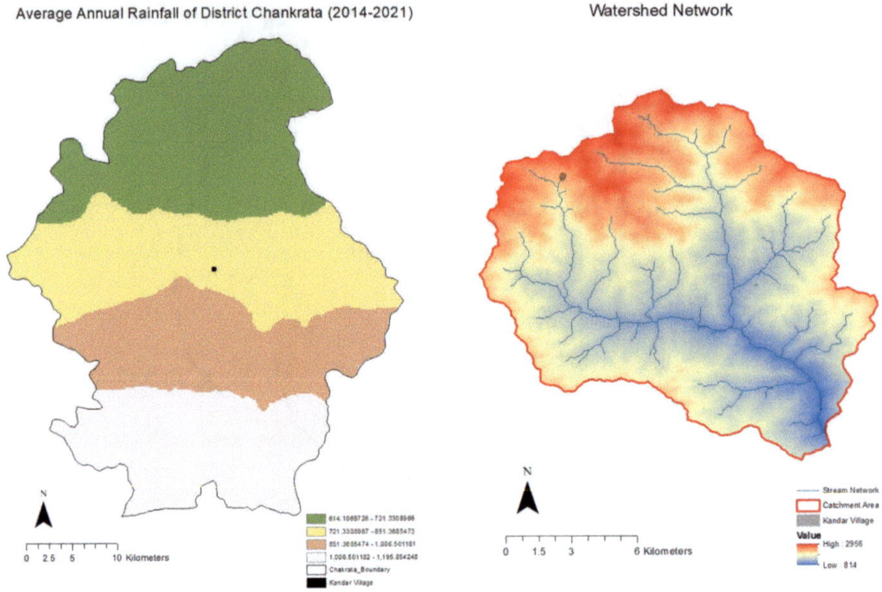

Fig. 3 Annual rainfall and watershed of Dehradun district (*Source* Author)

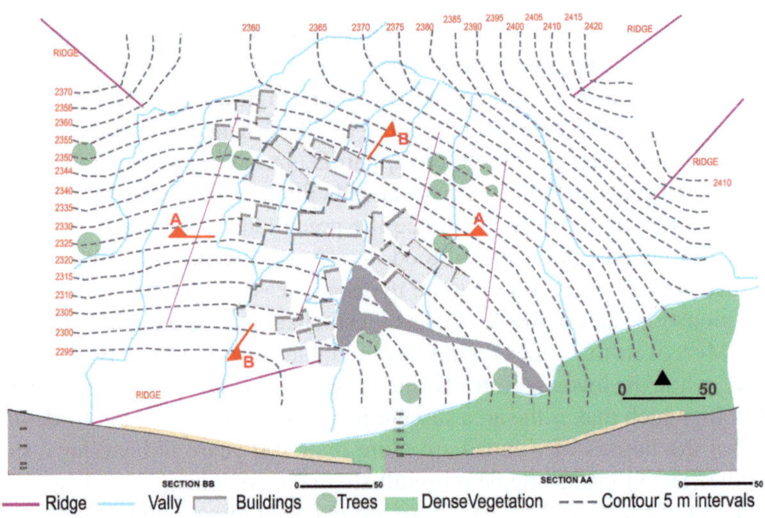

Fig. 4 Plan of Kandad village (*Source* Author)

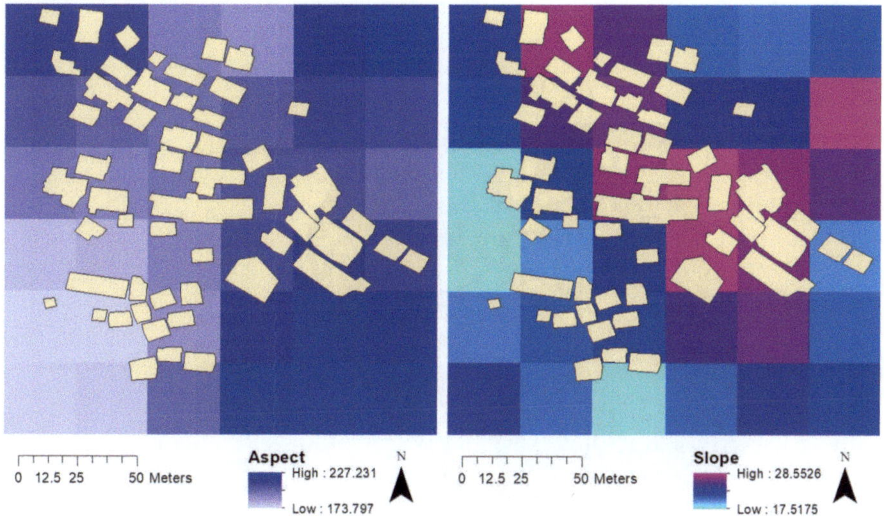

Fig. 5 Aspect and slope at Kandad village (*Source* Author)

undertake cattle rearing and farming for a specific time period at the upper level and then shift to the lower level during other times of the year with different cropping patterns and different cattle. Thus, indicating two sets of activities and habitation spaces on the upper and lower levels of the mountain terrain.

3.2 *Placement of the Settlements*

Figure 6 shows that the settlements are built in a long linear pattern to allow maximum sun. The buildings are placed along the contours in the stepped terraces allowing minimum disturbance to terrain. An analysis of the built-up change over time indicates the spread of the village is more toward the northern and southern edges along the contour lines (Fig. 7). The settlements located along the mountain regions and perhaps the maximum exposure to the sun direction throughout the year with a favorable combination of leeward side winds may suggest the settlement's location. It is natural that the "spread" of the form, its porosity, its insulation capacity, and the profile of the form all contribute to making internal conditions of living more comfortably. Hence, the snow doesn't accumulate on the roofs, or almost zero openings for reduced porosity are given to inhibit wind movement in the interiors.

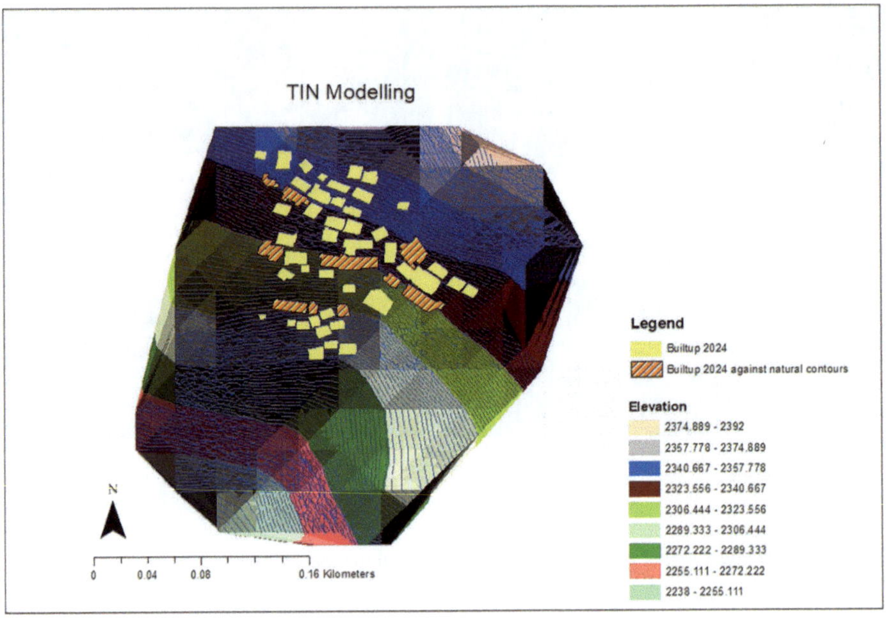

Fig. 6 TIN model of Kandad village (*Source* Author)

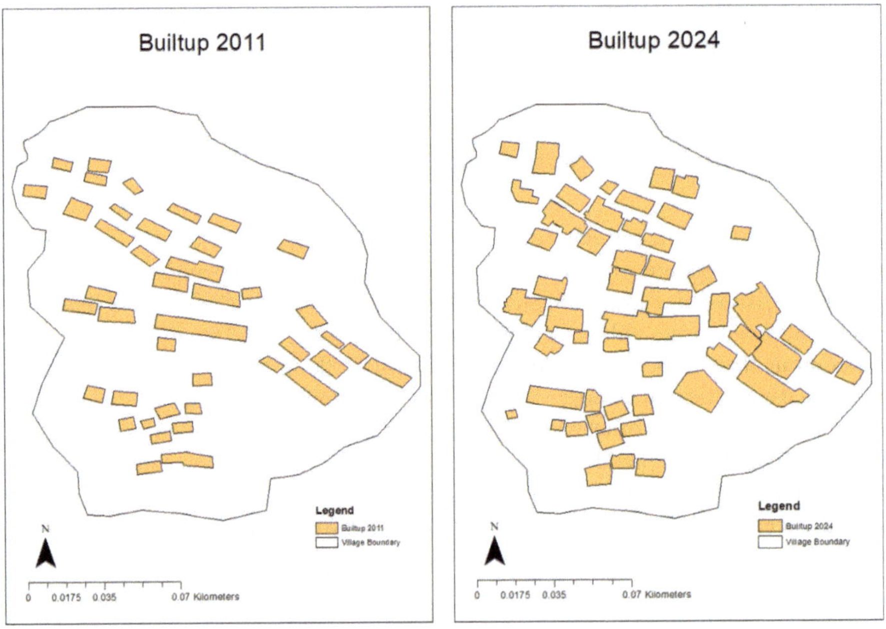

Fig. 7 Change in built-up in Kandad village (*Source* Author)

3.3 Response to Topography

Agricultural produce forms a large part of storage space for each family. Sometimes, the storage can go up to two floors with the topping of habitation spaces. The lower levels of storage are concealed or secured by a combination of thick stone walls and timber. The entire setup seems to keep the interiors dark and at constant temperature and also extremely secure from intruders. It also indicates that large storage spaces indicate a self-contained system that can last for almost four months of harsh winter. Thus, storage areas occupy almost 60–70 percent of the total built form, and the remaining 30 percent is given to habitation spaces, which are light in weight, made of timber and makes one feel warm and exposes habitation spaces to sunlight and larger distance is gained away from the cold lower terrain. It needs to be seen how constant cool temperatures assist in storages of various items of food and does that also help in maintaining hygiene of stored items on favorable says, when the ground is cleared of snow, the access to the lowest storage area is by an external door, indicating the ease of storing or stocking agricultural produce. During the harsh winters, when it is too unfavorable to expose oneself to external conditions, the entire structure acts as a container with vertical circulation ensured at all levels starting from upper habitation spaces to lower storage areas by means of internal ladders capped by floor lids. The partition walls at the upper-level double up as storage spaces—enclosing a cavity of 600 mm width and constructed of timber. Any of the small openings on roof or walls can be made to slide to allow a shaft of light to penetrate whenever required, else the interior feels like a cave.

Insulation is further enhanced by having thick walls constructed using a combination of local stone material and timber (local Deodar wood) at lower levels and complete timber construction at the topmost level of habitable spaces (Fig. 8). Stone walls are incorporated with timber beams used for structural purposes. In some places, Cob bricks, which are hand-made sun-dried bricks are used as load-bearing walls. Roofs are covered with locally available slates. Clay and dung are abundantly used because of their insulation properties. They help the rooms keep warm in winter and cool during the summer season. Mud paste is used for the plastering of walls (both outer and inner surfaces). Sloping roofs of the houses naturally drain the rainwater. Chajja over windows for shading and protection against rain. Floors are kept low so that natural warmth can be conserved. The height of doors and windows is kept very small for the same reason. The overall orientation of each form may also indicate maximum exposure to the sun (the longer side of the form facing sun exposure) and a shorter side perpendicular to wind movement. All construction materials are sourced from local areas, and it may invite the cooperation of the community for constructing the structures.

Depending on one's financial stability, one may choose to decorate or paint the interiors or alternatively keep it simple. Festivals, art symbols, decoration on fabric bespeak of the beliefs of the community. Decoding these symbols is a process demanding sustained venture to understand this phenomenon. The symbols may also offer clues regarding the imagination of their world of living.

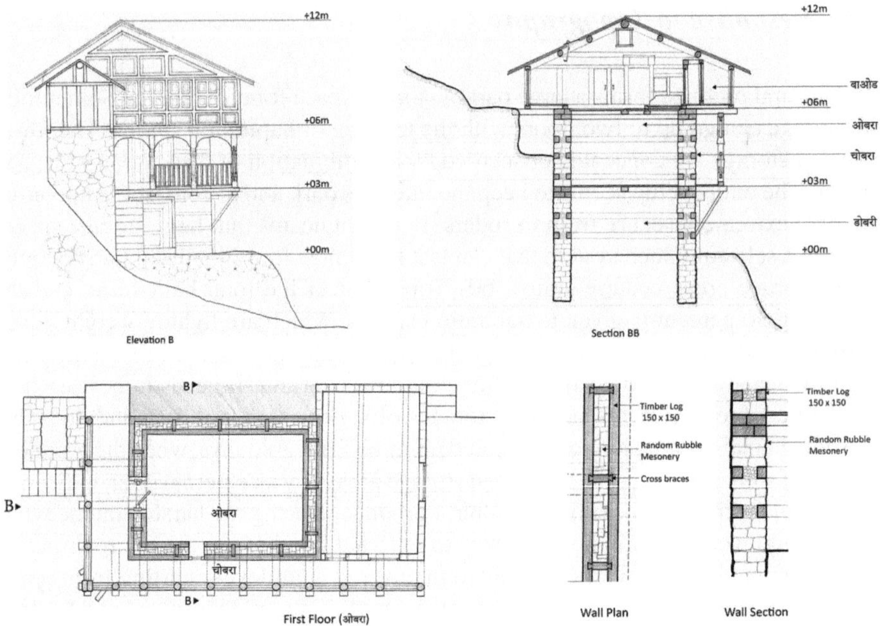

Fig. 8 Typical house in Kandad (*Source* Author)

3.4 Response to Culture

The living and cooking areas are of low height and are provided above the cattle space, fuel, and fodder space to provide warmth in winters. Approach to the living areas on the first floor is through the staircase on the side of the house. The balcony mostly 75cm in width in front of the house forms an integral part of the building. Construction materials like stone, wood, and slate are used extensively as they are locally available and easy to handle using manual tools, equipment, and manual labor. A wooden structural frame is made, and locally available stone is infilled for making the walls. The beams and columns made of timber are intricately carved to improve the aesthetics. Floors and ceiling of the building are also made from the wooden planks. The sloping roofs are made with slates and are supported over wooden trusses. The "Kothar" or the grain storage structure symbolizes the affluence of the family and is located near the house (Fig. 9). These are wooden structures placed 1m above ground level with a small gap all around the rectangular structure to keep it isolated so that the grains do not get damaged easily. Festivals, art symbols, decoration on fabric bespeak of the beliefs of the community. Decoding these symbols is a process demanding sustained venture to understand this phenomenon. The symbols may also offer clues regarding the imagination of their world of living.

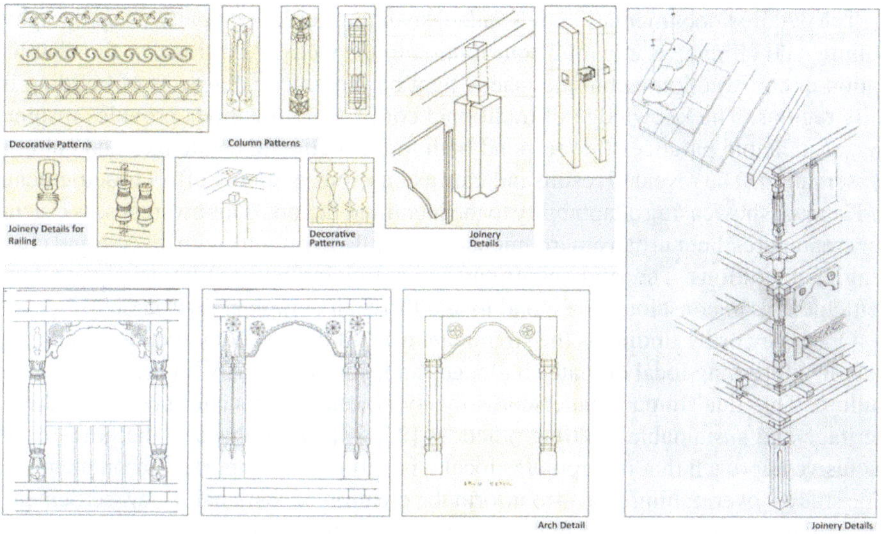

Fig. 9 Details of cultural symbolism and carvings on wooden structures (*Source* Author)

4 Discussion

The literature study underscores the pressing challenges faced by towns in hilly regions due to increased population density, necessitating thoughtful expansion solutions like the creation of satellite towns. The inherent difficulty in planning settlements in such topographies is acknowledged, emphasizing the importance of considering complex factors such as topography, geology, climate, and culture [1]. The study's primary objective is to unravel how existing settlements manage to strike a delicate balance between these critical elements. The initial phase of the research involves a meticulous documentation of the study area, spanning settlement patterns, geography, culture, and climate. This comprehensive approach aligns with the complexity of hilly regions and recognizes the interplay of multiple factors in shaping the urban fabric. Earlier research highlighted the need for similar approach [18, 19]. Additionally, the identification of construction techniques adds depth to the understanding, acknowledging the practical aspects of how these settlements are built within their challenging environments.

The utilization of geoinformatics in the subsequent analysis is a strategic move, providing a spatial lens to dissect the impact of identified factors. This technology-driven approach allows for a nuanced exploration of the intricate relationships between culture, climate, and geography in shaping the settlements. Earlier studies highlight the same observations [20, 21]. The revealed results, as outlined in the abstract, suggest a conscious and deliberate planning effort, indicating a responsiveness to the identified factors. This finding is pivotal as it showcases the adaptability of these settlements to their unique circumstances.

The detailed documentation of Kandad, spanning settlement patterns, geography, culture, and climate, is a pivotal contribution to the study. Additionally, the identification of construction techniques sheds light on the practical aspects of building in hilly regions. The knowledge of traditional construction methods provides insights into sustainable practices that respond both to local climate challenges and cultural preservation. The revealed results indicate a conscious planning effort in settlements of Kandad, showcasing adaptability to the identified factors. This awareness is crucial for creating resilient urban environments, especially in regions with diverse and challenging conditions. The study contributes to a growing body of knowledge on how settlements can consciously respond to and thrive in complex environments.

One noteworthy finding is the significant role traditional construction techniques play in addressing local climate challenges and, simultaneously, in preserving local culture. This dual impact underscores the symbiotic relationship between cultural heritage and sustainable building practices [22, 23]. It emphasizes the need for an inclusive approach that incorporates local wisdom in contemporary urban planning. The study's overarching goal—to inform the creation of sustainable new settlements in similar geographical and cultural conditions—bears practical implications for future development initiatives. The insights gained from Kandad can serve as a valuable guide for planners and policymakers grappling with similar challenges in hilly regions.

5 Conclusions

The study reveals that traditional construction techniques have been able to tackle the challenges of mountainous regions. This indicates that residents have a deep understanding of local environments and resource constraints. Hilly regions have rugged terrains, seismic activity, and extreme weather conditions prevail. In such a situation, traditional methods prioritize resilience, sustainability, and cultural relevance. Techniques such as stone masonry, timber framing, and earth construction are common, utilizing locally sourced materials that offer strength, durability, and thermal efficiency. The construction combination of stone, that offers stability against landslides and soil erosions and timber framing, allowing flexibility and seismic resilience, proves beneficial. Additionally, thick walls leverage thermal mass to regulate indoor temperatures, crucial in regions with fluctuating temperatures. Moreover, these methods incorporate in depth knowledge of site selection, orientation, and natural ventilation to optimize comfort and energy efficiency. By embracing traditional wisdom and indigenous practices, construction in hilly and mountainous region not only addresses the challenges posed by the environment but fosters a sustainable and culturally rich heritage.

In wake of new construction technology and rapid expansion of built, along with diminishing understanding of traditional knowledge systems, GIS can play important role in making sound decisions regarding planning of settlements. Further studies of similar settlements can help develop a comprehensive understanding regarding site suitability. It will help identify areas with stable terrain suitable for construction. The current study focused on documentation and did not go into the details of community participation aspects using new technology such as GIS for sustainable development. Hence, further studies can explore usage of GIS for community engagement and participatory decision-making. By involvement of local communities in the planning and implementation of traditional construction projects, GIS can help ensure cultural relevance and social acceptance.

Acknowledgements The documentation work is undertaken as part of Relative Study Program of VIT's Padmabhushan Dr. Vasantdada Patil College of Architecture's Second Year B. Arch. students in the Academic Year 2018-19.

References

1. Schickhoff U, Singh RB, Mal S (2022) Mountain landscapes in transition: effects of land use and climate change. Sustainable Development Goals Series. Springer. https://doi.org/10.1007/978-3-030-70238-0_23
2. Grêt-Regamey A, Weibel B (2020) Global assessment of mountain ecosystem services using earth observation data. Ecosyst Serv 46. https://doi.org/10.1016/j.ecoser.2020.101213
3. Tran VT, An-Vo DA, Cockfield G, Mushtaq S (2021) Assessing livelihood vulnerability of minority ethnic groups to climate change: a case study from the northwest mountainous regions of Vietnam. Sustainability (Switzerland) 13. https://doi.org/10.3390/su13137106
4. Dame J, Schmidt S, Müller J, Nüsser M (2019) Urbanisation and socio-ecological challenges in high mountain towns: insights from Leh (Ladakh), India. Landsc Urban Plan 189:189–199. https://doi.org/10.1016/j.landurbplan.2019.04.017
5. Roche PK, Campagne CS (2017) From ecosystem integrity to ecosystem condition: a continuity of concepts supporting different aspects of ecosystem sustainability. Curr Opin Environ Sustain 29:63–68. https://doi.org/10.1016/j.cosust.2017.12.009
6. Zuo Q, Zhou Y, Liu J (2022) Construction and optimization strategy of an ecological network in mountainous areas: a case study in southwestern Hubei Province, China. Int J Environ Res Public Health 19. https://doi.org/10.3390/ijerph19159582
7. Murton G (2017) Making mountain places into state spaces: infrastructure, consumption, and territorial practice in a Himalayan borderland. Ann Am Assoc Geogr 107:536–545. https://doi.org/10.1080/24694452.2016.1232616
8. Soriano MA, Diwa J, Herath S, De S, Province H, Aguilar-gómez CR, Franco-maass S (2016) Climate change impact in the western Himalaya : people's perception and adaptive strategies. J Mt Sci 14:1455–1472. https://doi.org/10.1007/s11629-015-3814-1
9. Livneh B, Badger AM (2020) Drought less predictable under declining future snowpack. Nat Clim Chang 10:452–458. https://doi.org/10.1038/s41558-020-0754-8
10. Olokeogun OS, Kumar M (2020) An indicator based approach for assessing the vulnerability of riparian ecosystem under the influence of urbanization in the Indian Himalayan city, Dehradun. Ecol Indic 119:106796. https://doi.org/10.1016/j.ecolind.2020.106796
11. Singh S, Tanvir Hassan SM, Hassan M, Bharti N (2020) Urbanisation and water insecurity in the Hindu Kush Himalaya: insights from Bangladesh, India, Nepal and Pakistan. Water Policy. 22:9–32. https://doi.org/10.2166/wp.2019.215
12. Patowary S, Sarma AK (2018) Model-based analysis of urban settlement process in eco-sensitive area of developing country: a study with special reference to hills of an Indian city. Environ Dev Sustain 20:1777–1795. https://doi.org/10.1007/s10668-017-9965-1

13. Patowary S, Sarma AK (2020) Projection of urban settlement in eco-sensitive hilly areas and its impact on peak runoff. Environ Dev Sustain 22:5833–5848. https://doi.org/10.1007/s10668-019-00453-x
14. Kumar S, Bansal VK (2021) GIS-based locational evaluation of infrastructure facilities in hilly regions: a case study of an institute campus. Int J Constr Manag 21:1165–1184. https://doi.org/10.1080/15623599.2019.1604114
15. Mallick J, Ibnatiq AA, Kahla NB, Alqadhi S, Singh VP, Hoa PV, Hang HT, Hong NV, Le HA (2022) GIS-based decision support system for safe and sustainable building construction site in a mountainous region. Sustainability (Switzerland) 14:1–33. https://doi.org/10.3390/su14020888
16. Kumar S, Bansal VK (2016) A GIS-based methodology for safe site selection of a building in a hilly region. Front Archit Res 5:39–51. https://doi.org/10.1016/j.foar.2016.01.001
17. Office of the Registrar General & Census Commissioner, India (2011) Census of India 2011: Primary Census Abstract, Series-1, India. New Delhi: ministry of Home Affairs, Government of India
18. Rus K, Kilar V, Koren D (2018) Resilience assessment of complex urban systems to natural disasters: a new literature review. Int J Disaster Risk Reduct 31:311–330. https://doi.org/10.1016/j.ijdrr.2018.05.015
19. Schirpke U, Candiago S, Egarter Vigl L, Jäger H, Labadini A, Marsoner T, Meisch C, Tasser E, Tappeiner U (2019) Integrating supply, flow and demand to enhance the understanding of interactions among multiple ecosystem services. Sci Total Environ 651:928–941. https://doi.org/10.1016/j.scitotenv.2018.09.235
20. Diksha, Kumar A (2017) Analysing urban sprawl and land consumption patterns in major capital cities in the Himalayan region using geoinformatics. Appl Geogr 89:112–123. https://doi.org/10.1016/j.apgeog.2017.10.010
21. Tenerelli P, Püffel C, Luque S (2017) Spatial assessment of aesthetic services in a complex mountain region: combining visual landscape properties with crowdsourced geographic information. Landsc Ecol 32:1097–1115. https://doi.org/10.1007/s10980-017-0498-7
22. Klein JA, Tucker CM, Steger CE, Nolin A, Reid R, Hopping KA, Yeh ET, Pradhan MS, Taber A, Molden D, Ghate R, Choudhury D, Alcántara-Ayala I, Lavorel S, Müller B, Grêt-Regamey A, Boone RB, Bourgeron P, Castellanos E, Chen X, Dong S, Keiler M, Seidl R, Thorn J, Yager K (2019) An integrated community and ecosystem-based approach to disaster risk reduction in mountain systems. Environ Sci Policy 94:143–152. https://doi.org/10.1016/j.envsci.2018.12.034
23. Pandey R, Jha SK, Alatalo JM, Archie KM, Gupta AK (2017) Sustainable livelihood framework-based indicators for assessing climate change vulnerability and adaptation for Himalayan communities. Ecol Indic 79:338–346. https://doi.org/10.1016/j.ecolind.2017.03.047

Accessibility for Senior Citizens in Hill Settlements: Challenges and Opportunities for Sustainable Mobility

Parshant Rehal, Prabhjot Singh Chani, Sonal Atreya, Apoorva Nawani, and Harsimran Kaur

1 Introduction

Most hill settlements, characterized by rugged terrain, steep slopes, and often remote locations, present unique challenges for residents, especially the older adult population. Accessibility in the neighbourhoods within and around the hilly areas can be particularly challenging due to geographical constraints, inadequate infrastructure, and limited resources. These conditions become even more acute for the ageing populations residing in hilly settlements, who often witness limiting physical and environmental barriers to their out-of-home daily activities. Therefore, this chapter explores the complex accessibility issue that older adult populations in hill settlements face, as well as the opportunities for sustainable mobility solutions.

P. Rehal · P. S. Chani · S. Atreya (✉)
Indian Institute of Technology, Roorkee, India
e-mail: sonal.atreya@ar.iitr.ac.in

P. Rehal
e-mail: prehal@ar.iitr.ac.in

P. S. Chani
e-mail: p.chani@ar.iitr.ac.in

A. Nawani · H. Kaur
Indian Institute of Technology (BHU), Varanasi, India
e-mail: apoorvanawani.rs.apd23@iitbhu.ac.in

H. Kaur
e-mail: harsimran.apd@iitbhu.ac.in

V. Chettry et al. (eds.), *Geoinformatics for Hilly Regions*, Advances in 21st Century Human Settlements, https://doi.org/10.1007/978-981-95-0156-4_6

1.1 Need for the Study

Several factors can be seen behind the need to address accessibility issues among older adults residing in urban hill settlements. The following sections try to identify the main reasons for the need to begin addressing this concern for older adults. The study also tries to explore existing literature resources to gain an understanding of a few concepts concerning relationships between older adults and their environmental needs and daily activity patterns.

Ageing Population and Future Demographic Trends

Many developed nations have been witnessing the demographic trends of population ageing, attributed to lower fertility rates and increased longevity. With the global adult population aged 60 years and above increasing from 11% in 2006 to 22% in 2050, this demographic trend will be further challenging for developing nations, including India, which will house more than 80% of the older adult population living in low- and middle-income groups by 2050, thus demanding more health and social systems. In India, it is estimated that the adult population aged 60 years and above will outnumber the most significant population groups.

Socio-Economic Conditions and Ageing Challenges

Urbanization and migration among younger adults have altered the age-old family and social structures, often causing social discord, leading to most older adults living a secluded and lonely life. Migration for work is particularly a widely evident trend among younger populations in hilly cities, which often fall behind in terms of employment opportunities due to a lack of industrial activities in such environmentally sensitive areas.

This demographic trend has led to an increase in nuclear family structures, creating a situation where most adult populations need to be functionally more independent and active despite declining physical abilities. Staying alone often further deteriorates life satisfaction, creating stress and dependence among older adults to fulfil life requirements in terms of health, social, and economic aspects. Such scenarios become critical and challenging for older adults, particularly those residing in urban hill settlements, as accumulating daily need commodities requires higher inputs of physical activities, including active mobility and the necessity for outdoor long-distance travel. On the contrary, older age is known to be associated with declining physical and cognitive abilities, a reduction in functional capacities, and increased tendencies for disabilities due to tripping and falling in outdoor built environments.

In wake of population ageing future projections, as most of the basic activities of daily life necessitate streamlined outdoor mobility and enhanced accessibility in outdoor environments, it becomes critical to enhance the research quotient towards addressing the accessibility needs of older adults to help invigorate inclusiveness, equity of opportunities, and enhanced social connectedness, and thus help in making our demographic futures more age-friendly. It also becomes imperative to plan for an age-friendly built environment that may be capable of eliminating obstructions to outdoor functional abilities without the fear of falling or tripping, thereby diminishing the need for age care and healthcare infrastructure for healthy ageing.

1.2 Literature Review

Researchers have exhibited the effects of outdoor built environments in neighbourhoods and cities on the quality of life and well-being of older adult populations. The liveability of the living environment among ageing populations is often seen as an opportunity for enhancing functional independence, daily social participation, and building social infrastructure and amenities [1]. Thus, it becomes critical to study and comprehend person-environment fit from the concepts of active ageing and ageing-in-place as the vital processes for experiencing higher liveability and well-being among older adults.

Active Ageing

Active ageing is often referred to as an opportunity for participation and inclusion in different social, economic, spiritual and civic activities, along with maintaining physical and mental health [2]. Many secondary literature resources exhibit the effects of active ageing and the availability of green open spaces on delays in health concerns and physical ailments arising from the immediate effect of biological ageing among human beings.

There is an abundance of research which exhibits the importance of the availability and quality of green open spaces around the living environments. Researchers have extensively linked the potential for active ageing to the availability and quality of open spaces, which may boost recommended levels of weekly physical workout time [3]. While such planned green and open spaces are in abundance in settlements with a flatter topography, examples of planned green open spaces in the urban hill settlements are quite rare. The areas exhibiting deprivation trends in such social infrastructure pose many challenges to outdoor activity among elderly people [4].

Instead of the planned green open spaces in hill settlements, while neighbourhood streets are primarily seen as mere walking spaces adjoining the dwelling areas, but such streets are also distinguished as places of active engagement and active living among older populations. Pedestrian roads and streets in hill settlements having positive characteristics such as restricted vehicular entries, aesthetically pleasing design and pleasant weather provide higher safety and comfort for pedestrian users, thus proving to be highly liveable and desired spaces for physical and social engagements among older adults [5].

Ageing-in-Place

Ageing-in-place is perceived as insidedness with the neighbourhood and cities one is usually settled in over time, which is enough to familiarize one with the environment and living conditions [6]. Extended periods of stay and familiarity within the neighbourhood and city influence older adult individuals to live independently and maintain social connectedness and participation. Fear of losing familiarity and social connection is usually seen as a retardant and prohibitive factor among older persons when choosing to move to newer avenues due to the migration of younger family members to other cities. Researchers have linked psychological depressive symptoms in geriatric groups to levels of adaptation abilities to the built environment conditions and ageing-in-place [7].

In addition to the person-environment fit, Wennberg et al. (2010) highlight the rightful usability of the available infrastructure in removing the barriers to accessibility [8]. While physical frailty often becomes the primary reason behind reluctance to visit out-of-home environments for leisure, recreation, or physical activity [9], fears of falling or tripping in outdoor environments are enhanced due to the poor quality of pedestrians and sidewalks.

Particularly in urban hill settlements where streets present an array of accessibility concerns, a larger share of older adult populations are subjected to staying indoors, leading to social disengagement. The deprivation caused by the lack of quality-built environmental conditions can impair and limit outdoor activity among elderly people, affecting their functional independence [10]. Steinfeld also considers that while cultures and professional values in different geographical areas have varying implications, improving accessibility conditions can be vital to encouraging older adults to be functionally active, connected, and independent.

Mobility and Accessibility in Hill Settlement

While addressing functional independence among older adults, out-of-home mobility becomes a critical perspective in developing sustainable built environments. Researchers are increasingly focusing on the mobility and accessibility of neighbourhoods and cities while investigating active ageing and the perception of enhanced ageing-in-place. Mobility has often been perceived as the ability to travel across desired distances, ranging from the home environment to regions beyond, and encompasses independent, assisted movements and transportation. While the significance of utilizing public or personal vehicle options for long-distance travel is universally acceptable, active mobility, particularly walkability, is predominantly perceived as a determining factor for the enhanced functional capabilities of older adults in hill settlements. When seen in the context of hill settlements, where active mobility and walkability are often the prominent concerns for functional independence, a lot depends on how planning strategies and design improve the accessibility conditions. Active mobility in many urban hill settlements in India is a critical challenge for the older adult populations to perform their activities of daily life (ADLs).

Accessibility is often referred to as a function of proximity to services and basic life amenities and places, and it has been often studied in terms of physical and non-physical characteristics. Researchers have discussed accessibility in terms of physical factors (availability, proximity, and route quality) and non-physical factors encompassing adequacy, acceptability, and affordability [11].

While research from the developed nations highlights concerns and initiatives that have been successfully implemented towards building age-friendly neighbourhood and city environments, there has been a paucity of comparable research and initiatives in India, especially when seen in the context of developing sustainable age-friendly solutions in the context of hill settlements characterized by accessibility concerns due to steeper terrain.

The research identifies mobility and accessibility as a primary concern in building an inclusive, sustainable built environment, community, and environment for older adults in hill settlements. The study attempts to examine the various accessibility challenges posed by the geographical constraints, inadequate infrastructure, and inadequacy of design standards prevalent in hill settlements, which hamper the abilities of older adult residents residing in such locations.

2 Methodology

Building upon the understanding gathered through a literature study concerning out-of-home engagements among older adults and the mobility and accessibility of places of daily outdoor activities, the study attributes to identify different challenges and barriers for mobility and accessibility experienced by older adults in hill settlements. The study gathers information about the barriers to accessibility and comfortable outdoor mobility through observational analysis and reported concerns of older adults. Challenges related to accessibility arising from geographical constraints, infrastructural deficiencies, transportational conditions, inadequacy of design standards, and concerns of older persons related to the inadequacy of open space and their accessibility concerns are identified in the following section, employing practical examples from a few Indian hill cities. Primary imagery data and imagery from Google Street views have been used from a few hill towns in Himachal Pradesh and Uttarakhand, India. Most of these towns included in the study are undergoing infrastructural changes under the Smart City Mission in India. But despite these initiatives, the existing conditions fail to address the mobility and accessibility needs of older adults.

3 Accessibility Challenges

Hill settlements have, though, seen a significant thrust in improving intercity connectivity through the expansion of road networks in India. Improving accessibility and walkability and improving universal accessibility at the neighbourhood level have not received ample attention in most Indian hill towns and cities. Such conditions may not be seen as threatening by young adults, but these conditions become both physical and psychological barriers for older adults to carry out their usual outdoor activities comfortably.

The following text identifies various accessibility challenges that hamper the quality of life among older adults concerning geographical constraints, infrastructural deficiencies, and healthcare accessibility.

3.1 Geographic Constraints

Most hill settlements pose numerous challenges for older people and are almost uninviting for older people. Frail and sensitive environments, paucity of developable land, and steep topography are often among the various geographical constraints that affect planning adequate pedestrian and street networks that are accessible for older adults and persons with disabilities. Expansion of the hill settlements outwards often increases the distance to reach basic life necessities and the performance of ADLs, thus exerting more dependence on vehicular activities and limiting the chances to choose active and sustainable modes of travel. Steep slopes and winding roads increase the walking distance, making walkability an unviable option for older adults. In most cases, sidewalks are absent in most areas, leading to most people walking on roads, posing a threat to their safety. While uphill access through convenient ramps is not provided, even access through staircases becomes inconvenient for older adults and children due to uneven riser heights and tread widths (Fig. 1).

Older adults are often vulnerable to falls and tripping accidents while traversing down or upwards walkways with steep slopes, and the absence of railing on both sides makes it further difficult for older adults and kids to walk comfortably to their destinations. Railings, when present, are often discontinuous and poorly designed (Fig. 2).

Fig. 1 Absence of sidewalks, ramps and staircases with non-uniform dimensions near the Old Bus Stand Area, Shimla

Fig. 2 **a**, **b** Steep walkways without railing or grab rails **c** discontinuous and obstructed railings. *Source* Author

3.2 Infrastructure Deficiencies

Upgrading the infrastructural deficiencies does not always follow the same envisioned pace of urbanization when compared to urban areas in plain topography due to an array of reasons, including a higher cost of construction and a lack of developable land. The demand for development in hill settlements in urban areas has remained high under the stresses of the population. However, the same is unviable due to the sensitive and fragile environments surrounding the hill settlements which are vulnerable to man-made and natural calamities due to rapid and unscientific approaches of development [12]. Most hilly areas witness higher tendencies of natural calamities like cloudbursts, earthquakes, and flash floods, further damaging the available infrastructure for roads, public transport, and housing. Heavy rains and snowfall often damage the road and street surfaces, making the conditions vulnerable for older adults to carry out active trips between home and other destinations.

3.3 Transportation Conditions

While most residents in hill settlements heavily engage in walkability in local-scale environments, the same may not always be suitable for the ageing populations who face diminishing physical abilities and gait-related issues, forcing most elderly persons to confine themselves to their homes and thus increasing their dependence. Studies exhibit an apparent decline in the driving capabilities of older adults due to varying health conditions and lower motor and cognitive skills, despite personal vehicles being the most convenient mode of travel [13]. Even with driving abilities, driving becomes stressful through narrow hill roads with high traffic volumes and searching for parking spaces around different destinations.

3.4 *Inadequacy of Design Standards*

When hill settlements are concerned, there is a paucity of available guidelines and strategies for pedestrian accessibility for older adults, which are mainly missing in contrast to the case of plains. Existing design standards often fail to be an effective way of planning accessible pedestrian infrastructure on hills. Most pedestrian and sidewalk networks between neighbourhoods and other public areas are often built inadequately, undermining the importance of universal accessibility standards (Fig. 3). Poor universal accessibility is often evident at most outdoor destinations in hilly neighbourhoods and cities. The absence of lifts or provisions for ramps is often one of the factors that make outdoor destinations inaccessible for older adults who use walking aids (Fig. 4).

Fig. 3 Steep and non-continuous pedestrian sidewalks. *Source* Author

Fig. 4 Poor universal accessibility measures in public buildings. *Source* Author

Fig. 5 Poorly built pedestrian facilities in hills along road networks. *Source* Author

At most locations, pedestrians and sidewalks are poorly built and ill-maintained and do not meet the design standards (Fig. 5). Inefficient planning policies add further woes to the satisfaction levels among older adults concerning accessibility conditions. Context and site-specific accessibility guidelines are absent for implementation during the design and development stages. With steeply designed uneven steps and the absence of provisions for ramps with comfortable walking slopes, most basic places like retail facilities, shops, ATMs, bus stops, hospitals, parks, other open spaces, etc. often become inaccessible to frail and disabled older persons. In most public places, steep steps and the absence of railings often subject older people to fatalities like tripping and falling, causing severe injuries.

Poorly designed walking surfaces on sidewalks along roads often force residents to walk on the streets. In most Indian hill cities, pedestrian infrastructure along roads is poorly built, and there are hardly any resting spaces or pedestrian amenities like clean drinking water, dustbins, or adequate washroom facilities. In most hilly cities, finding pedestrian signals, designed zebra crossings, and speed control devices is rare. Providing disabled-friendly transport options has still not gained attention in most Indian cities, including the hill settlements.

3.5 *Open Spaces Accessibility and Inadequacy*

Most Indian hill cities have a lot of natural open areas in the form of reserve forests or undeveloped natural sites. However, when the availability of accessible man-made green open spaces is considered, most neighbourhoods in hill settlements fail to provide such areas adequately, which are otherwise necessary for enhancing active lifestyles and encouraging social and intergenerational connectivity. Most parks, when available in the case of hills, have walking paths that are often steeply sloped and are not accessible for older persons with independent or assisted walking conditions. The lack of amenities, improper lighting, and refraining access to stray animals often render these spaces unsuitable and unsafe for older adults. Most of the time, such man-made spaces are absent within various localities, thus limiting the tendency for comfortable walking connectivity to these open spaces.

4 Towards Building Sustainable Age-Friendly Communities and Cities

The World Health Organization prioritizes different aspects related to infrastructural factors like housing, transportation, outdoor built environments, health determinants, and social inclusion and participation as the fundamental pillars for building age-friendly cities and urban areas [14]. The United Nations has declared the 'Decade of Healthy ageing' in this contemporary decade (2021–30). While opportunities for active ageing have apparent effects on physical health, they also help maintain mental health and well-being.

In order to enhance the opportunities for active ageing, age-friendly environments play a significant role by providing built environments that foster functional abilities among older adults. While the out-of-home environments within the neighbourhood environments within the hill settlements can be primarily challenging, demanding relatively higher physical efforts from the older adults, efforts to introduce urban design initiatives and create barrier-accessible environments can immensely create an age-friendly environment in the hill settlements.

The principles of universal design act as guiding principles in developing products and environments that provide simplistic yet innovative equity and flexibility of use for people with varying health and physical conditions. The principles of making the products, infrastructure, and environment with appropriate size, fit, and space, which provide perceptible information despite varying sensory and cognitive abilities, provide tolerance for errors, and require low physical efforts, not only help the persons who experience any disability but also the universal population to a much larger extent. URDPFI lays guidelines on various aspects of building out-of-home environments, which WHO considers in the direction of building age-friendly cities.

The following section explores different aspects of age-friendly neighbourhood and city designs in the following text:

4.1 Providing Green and Open Spaces in Hill Regions

Open spaces around the housing environment within the urban neighbourhood promote active ageing and social connectedness opportunities and generating intergenerational mutual care in society, also providing opportunities for pollution-free built environments within otherwise congested urban areas. Broadly categorized into recreational, organized green, and open spaces, the space requirement for open spaces, specifically for hills, must consider protected zones and ecological conservation areas over and above the 10–12 sq. m per person as suggested by URDPFI guidelines. In contrast, NBC suggests an area of a minimum of 3 sq. m per person, excluding recreational spaces, vacant land, flood plains, and forests. Also, the standards prescribe the minimum area requirement for the hierarchies of organized and man-made greens, which can be planned at the housing clusters, neighbourhoods, community, district, or sub-city levels as per NBC 2005 as visible in Table 1, where it is mandated to create a housing level park for a population of 5000 persons and a neighbourhood level park for a minimum of 10,000 persons, along with the creation of a botanical garden for every town and a recreational complex for every settlement, which inherits the potential for tourist attractions [15].

While specifically designing and creating man-made greens within the neighbourhoods of hill settlements, most of such places are located on varying contour levels. Thus, providing staircases with uniform and low-height risers and ramps with optimum slopes becomes imperative to assist older persons, including disabled persons, to traverse the level difference comfortably. While railings on both ends become necessary, intermediate landings should be considered. Wherever possible, adequate walking and sitting spaces should be provided on relatively flat land to not compromise the concerns of specially abled people. Older adults who use walking aids or an assisted wheelchair should have easy access through the entrance to such spaces. Considering the physical abilities of older adults, it is also necessary to consider the comfortable range of walking, thus making it necessary to provide shaded seating spaces adjoining the walking stretches (Fig. 6).

Table 1 Organized greens for hilly areas. [15]

Sr	Category	Population per unit	Area requirement in hectares
1	Housing area park	5000	0.50 to 1.00
2	Neighbourhood park	10,000	1.2 to 2.00
3	City parks/playgrounds/ maidans/ exhibitions grounds/cultural gathering grounds	For the entire town at different sites depending upon the availability of land	–
4	Botanical park	1 for every town	10.00 to 20.00
5	Recreational complex including zoo	1 for every town based on tourist potential	10.0 to 12.00

Fig. 6 Seatings for aged persons along usual trip locations. *Source* Author

At the same time, the width of walking paths inside such green parks should provide adequate walking space for at least two wheelchairs to cross comfortably. Given the colder weather conditions in most hill settlements, public sun-basking spaces can be planned along the walkways, and it is ideal for creating parklets and well-designed seating spaces along the streets and paths joining the home and the outdoor open spaces.

Providing basic amenities like clean potable water sources and public toilets, which are disabled-friendly, helps foster the desire for an active lifestyle, encouraging them to walk to carry out basic life activities outdoors.

4.2 Design Standards for Accessibility in Hill Settlements

Right of Way (ROW) becomes the crucial factor while planning and designing roads in the hilly terrains. Elements like carriageway, shoulder, side drain, etc. are the few important elements which are required in a road built in the hills. Standards for such road and pathway infrastructure are usually uncontrolled owing to varying slope steepness and availability of land, thus necessitating case sensitive design. Desirable widths of road land (ROW) for various categories of roads are given in Table 2.

The Town and Country planning, Himachal Pradesh thus prescribes the width of road in hilly regions in accordance with IRC:52-2019 and URDPFI guidelines 2014, which can be adopted as required for different land uses (Table 3).

While there are design standards for roads for both urban and hilly terrain, there are no explicit design standards for pedestrian facilities for hilly regions such as

Table 2 Desirable road land widths (in metres) [16]

Sr	Road classification	Open areas		Built-up areas	
		Normal	Exceptional	Normal	Exceptional
1	National and state highway (double lane)	24	18	20	18
2	Major district roads	18	15	15	12
3	Other district roads	15	12	12	9
4	Village roads	9	9	9	9

Table 3 Norms for carriage way in hills [17]

1-Single lane without kerb	05.00m
2-Lane without kerb/with kerb	08.80m/10.00m
3-Lane without kerb/ with kerb	11.80m/13.00m

width of footpath, kerb design, etc. The current standards for pedestrian facilities are mainly denoted by IRC:103-2012. The recommended design standards prescribing the width of footpath and ramp details are provided in Tables 4 and 5 respectively. A standard kerb ramp at least 1200 mm is integrated into the footpath and flushed with roadway at a gradient not greater than 1:12 and flared sides providing transition in three directions.

Table 4 Required width of footpath as per adjacent land use [18]

Sr	Landuse classification	Width (m)
1	Minimum obstacle free walkway width and residential/mixed use areas	1.8
2	Commercial/mixed use areas	2.5
3	Shopping frontages	3.5–4.5
4	Bus stops	3
5	High-intensity commercial areas	4

Table 5 Minimum specifications for walkable ramps [18]

Sr	Level difference (mm)	Minimum gradient of ramp	Ramp width (mm)	Handrails both sides	Comments
1	≥150 ≤300	1:12	1200	20	–
2	≥300 ≤750	1:12	1500	15	Landings every 5 m
3	≥750 ≤3000	1:15	1800	12	Landings every 9 m
4	≥3000	1:20	1800	9	Landings every 9 m

Fig. 7 Provision of safe road crossing designs, proper signages, pedestrian signals, and information panels. *Source* Author

4.3 Barrier-Free Mobility and Accessibility

Creating an environment devoid of barriers and emphasizing mobility that prioritizes accessibility can contribute to the development of an all-encompassing and sustainable community for the elderly. Primarily in hill settlements where accessibility conditions for older adult populations are comparatively poor, there is a greater opportunity and need to work in this direction in anticipation of the demographic trends of India's aging population over the next few decades. Establishing barrier-free and walkable connections between the residences and locations frequented by senior individuals facilitates their engagement in outdoor pursuits in a comfortable manner, thereby enhancing their level of functional autonomy.

Smooth, nonslip walkways and footpaths along roads and streets with vehicular access enhance the sense of security among the elderly. The implementation of secure buffer zones, such as vegetative flowerbeds or green parklets accompanied by bollards, effectively reduces the likelihood of collisions between pedestrians and vehicular traffic.

Traffic-calming devices such as speed bumps and tables, pedestrian underpasses at congested road crossings, and road crossings with median islands are all effective strategies for mitigating vehicle velocities. The integration of auditory and visual pedestrian signals at intersections offers a comprehensible resolution that caters to all demographics, including the elderly. Legible signage on roads and street surfaces, in addition to traffic signals, aids in reducing confusion among the elderly (Fig. 7).

4.4 Public Transportation Choices

It is necessary to plan for age-friendly, affordable, and accessible public transportation to create sustainable communities and cities. While hills pose a typically different challenge from the plains, where wide enough road networks are geographically not viable to link the usual activity areas with homes for the elderly extensively, solutions like call taxis, ropeways, and age-friendly smaller means for public transport are

Fig. 8 Low floor public transport. *Source* Author

effective for providing service to older adult populations, thus providing a necessary linkage for health, social connectedness, and inclusion.

Available, affordable, reliable, age-friendly, and disabled-friendly transportation choices are a few aspects that can help engage older and older adult populations in daily life out-of-home activities independently and maintain the desired level of social connection. Concessional travel passes that prioritize seats for older adults and disabled persons in all sorts of public transport are already prevalent in India. Some prevalent technologies for assisted travel for older adults, like low-floor vehicles with mechanical systems to assist wheelchair users in getting into lockable spaces within buses, trains, and other public vehicles, should be incorporated into newer transportation choices (Fig. 8).

Providing prioritized older adults and disabled-friendly reserved parking spaces adjoining essential service destinations is also essential. Above all, strict adherence to traffic rules and prioritizing the right of way for pedestrians is the most important quality that needs to be aroused among the Indian population.

5 Conclusion

The chapter highlights the accessibility challenges experienced by older adult populations towards their independent functioning capabilities, social connectedness, and well-being in the hilly settlements and along with, also discusses spatial strategies to enhance the age-friendliness of urban hill settlements. While the disparities related to economic and health aspects remain distinct for elderly populations, it is the opportunity for accessibility in out-of-home environments that can promote inclusiveness and equity of opportunities among older adult populations. It becomes imperative to include older adults and people with disabilities in planning procedures while designing and retrofitting urban environments. Enabling environments for older people will become a boosting factor to enhance the participation of this important demographic group. While most hill settlements target expanding tourism activities as an economic resource, focusing on accessibility and creating barrier-free environments can provide a much-needed boost. Improving accessibility of our hill settlements concerning older adult populations will be an enabling factor for this demographic group and thus help to participate in the process of building sustainable communities and cities.

References

1. Heatwole Shank KS, Cutchin MP (2016) Processes of developing 'community livability' in older age. J Aging Stud 39:66–72
2. WHO (2002a) Active ageing: a policy framework, vol 11. Accessed 2 June 2020
3. Yung EHK, Conejos S, Chan EHW (2016) Social needs of the elderly and active aging in public open spaces in urban renewal. Cities
4. Sugiyama T, Ward Thompson C (2008) Associations between characteristics of neighbourhood open space and older people's walking. Urban For Urban Green 7(1):41–51
5. Wang Z, Zhang H, Yang X, Li G (2022) Neighborhood streets as places of older adults' active travel and social interaction—a study in Daokou ancient town. J Transp Health 24:101309
6. Woolrych R et al (2020) Ageing in urban neighbourhoods: exploring place insideness amongst older adults in India, Brazil and the United Kingdom. Psychol Dev Soc 32(2):201–223
7. Yıldız M, Varol E, Elkoca A, Yildirim MS, Sarpdağı Y (2023) Determining the correlation between old age adaptation, aging in place and depression in older adults: a descriptive correlational study. Arch Psychiatr Nurs 43:153–161
8. Wennberg H, Hydén C, Ståhl A (2010) Barrier-free outdoor environments: older peoples' perceptions before and after implementation of legislative directives. Transp Policy 17(6):464–474
9. Chen M, Bolt G, Yu L, Hooimeijer P (2023) The impact of the residential environment on Chinese older people's ageing-in-place intentions: a mediation and moderation analysis. Habitat Int 140:102908
10. Steinfeld E, Danford GS (1999) Enabling environments: measuring the impact of environment on disability and rehabilitation. Springer Series in Rehabilitation and Health (SSRH)
11. Zandieh R, Martinez J, Flacke J, Maarseveen MV (2015) Inequality in access to local facilities and older adults' walking behaviour: an environmental justice perspective. In: Proceedings of the 52th international making cities livable conference on achieving green, healthy cities, vol 92, pp 1–15
12. Kumar A, Pushplata (2013) Vernacular practices: as a basis for formulating building regulations for hilly areas. Int J Sustain Built Environ 2(2):183–192
13. Falkenstein M, Karthaus M, Brüne-Cohrs U (2020) Age-related diseases and riving Safety. Geriatrics 5(4):1–28
14. World Health Organization [WHO] (2007) Global age-friendly cities: a guide. World Health Organization; Geneva, Switzerland
15. Bureau of Indian Standards [BIS] (2005) National building code of India 2005. Bureau of Indian Standards; New Delhi, India

16. Indian Road Congress [IRC] (2019) IRC:52-2019 guidelines for the alignment survey and geometric design of hill roads. Indian Road Congress; New Delhi, India
17. Government of India Ministry of urban affairs [GOI, MOUD] (2014) Urban and regional development plans formulation and implementation (URDPFI) guidelines (vol 2). Government of India ministry of urban affairs, New Delhi, India
18. Indian Road Congress [IRC] (2012) IRC: 103-2012 guidelines for pedestrian facilities. Indian Road Congress, New Delhi, India

Assessment of Urban Growth Dynamics in Srinagar (J&K), India

Sneha, Vishal Chettry, and Avinash Yadav

1 Introduction

Globally, urbanization has emerged as a significant driving force behind environmental change, markedly altering hill towns [1]. It is especially evident in developing countries where the pace of urban growth has been rapid, but predominantly unplanned, and unregulated. Hill towns are usually located in tectonically active and densely populated regions and have seen a surge in urban expansion over the past few decades [2]. Such a pattern is observed in previously less accessible areas, mainly due to improved road connectivity, promotion of new tourist attractions, progress in horticulture, the broadening of economic horizons, and a shift from primary sectors to secondary and tertiary ones, worsened by the absence of urban land use regulations [3]. These hill towns located across the world have witnessed a significant rise in the magnitude, extent, count, and intricacy of urban settlements, leading to a broadening of urban land use. Rapid urbanization has disrupted local hydrological regimes, lowered groundwater recharge, and reduced water availability for essential needs, marking a detrimental impact on fragile ecosystems of hilly regions [4]. It has also escalated the risks associated with natural hazards and increased the vulnerability of hill town residents to water, food, livelihood, and health insecurities. Climate change further strains these urban ecosystems by increasing the occurrence of extreme weather events [6]. The necessity for effective land use policies and integrated urban–rural planning is emphasized to guide urban growth in hill towns toward a more sustainable path, ensuring the conservation of essential natural resources.

Sneha · V. Chettry (✉) · A. Yadav
Department of Architecture, Planning and Design, IIT(BHU) Varanasi, Varanasi, India
e-mail: vishal.apd@iitbhu.ac.in

Sneha
e-mail: sneha.apd@iitbhu.ac.in

A. Yadav
e-mail: avinashyadav.rs.apd23@itbhu.ac.in

V. Chettry et al. (eds.), *Geoinformatics for Hilly Regions*, Advances in 21st Century Human Settlements, https://doi.org/10.1007/978-981-95-0156-4_7

Hill towns in India occupy a prominent position due to their unique geographical location and topographical features [5]. These towns are located in the mountain ranges that spans across the subcontinent and often on the edge of varying altitudes, ranging from the snow-clad peaks of the Himalayas in the north to the lush slopes of the Western and Eastern Ghats in the south [6]. This elevation differential not only provides respite from the scorching heat of the Indian plains but also results in a diverse range of ecosystems, making these towns critical biodiversity hotspots [7]. Moreover, their proximity to perennial rivers and abundant natural resources has historically attracted settlement, agriculture, and trade. The geographical significance of hill towns extends beyond their environmental richness. Many of these towns are strategically located along ancient trade routes, acting as gateways to remote regions and fostering cultural exchange. Furthermore, the challenging terrain has often necessitated innovative engineering solutions in the construction of roads, bridges, and settlements.

In recent years, rapid urbanization in hill towns of India has triggered a range of environmental challenges which includes but are not limited to deforestation, soil erosion, and habitat degradation, as pristine forest areas are cleared for infrastructure project. Further, the haphazard urban growth has contributed to a loss of biodiversity and disrupts the delicate ecological balance of these regions [8]. Moreover, a rise in the influx of tourists has led to an increase in water demand, air pollution from vehicular traffic and improper disposal of wastes. Inadequate waste management systems strain the local environment, and the demands for water resources for tourism can lead to over-extraction, depleting natural water sources and causing water scarcity in these ecologically sensitive areas. Therefore, balancing the need for urban development with the preservation of the fragile ecosystems of hill towns is a pressing environmental concern in India [9].

The geographical and environmental studies related to hill towns in India form a multidisciplinary field of research that offers crucial insights into the unique landscapes and ecological systems of these elevated regions. Researchers had delved into the geological processes shaping these terrains, studying the formation of mountain ranges, identifying geoheritage, and promoting ecotourism [10]. Several geological locations within Leh region in India hold significance in unraveling the processes behind the formation of the Indian subcontinent, the development of the formidable Himalayan mountain range, the history of the Tethys Ocean, as well as the past tectonic and climatic conditions of the area [11]. Climate and weather patterns are meticulously examined to understand the distinctive climatic conditions associated with varying altitudes and its impact on the availability of water and other natural resources [12]. Investigations into glaciers provide vital data for assessing water resource sustainability, especially in regions like the Indian Himalayan Region [13]. Additionally, biodiversity studies in hilly regions highlight the exotic flora and fauna found in these regions and orients the focus toward the endemic species and driving conservation efforts [14]. Research in this domain also delves into the natural hazards faced by the hill towns, such as landslides and flash floods, offering valuable insights for hazard mitigation [15].

Studies have also examined the historical evolution of hill towns, tracking urbanization trends and their impacts on local communities [16]. Socio-economic aspects are a significant area of focus, with researchers examining employment opportunities, education, healthcare infrastructure, and the implications of tourism industry on local economies and social structures. Additionally, impact of urbanization on the land degradation, deforestation, waste management, and water scarcity, along with the efforts toward sustainable development and environmental conservation in these towns are also studied [17]. Urban planning and vernacular architectural strategies also emerge as crucial themes, with research analyzing how design guidelines and planning practices contribute to sustainable cities and communities [18]. These myriad investigations offer an understanding of the diverse challenges and opportunities present in hill towns of India, thus contributing to the formulation of informed strategies for their sustainable development.

The Sustainable Development Goals (SDGs) outlined by the United Nations are especially pertinent in the context of hill towns, where unique geographical and ecological challenges intersect with human development [19]. The hill towns, with their fragile ecosystems, require focused sustainable development strategies to ensure the well-being of their communities while preserving the natural environment [20]. The emphasis of SDG 11 to promote sustainable cities and communities can be pivotal in hill towns where urban planning needs to account for steep terrains, risk of landslides, and limited accessibility. Similarly, SDG 6 concerning clean water and sanitation, and SDG 15 pertaining to life on land, are crucial as they underline the importance of safeguarding water resources and promoting sustainable use of terrestrial ecosystems, respectively, which are often at risk due to urban expansion and increased tourism. Through holistic planning that encompasses enhanced infrastructure, promotion of local culture, sustainable tourism, and community engagement, hill towns can stride toward achieving the SDGs while upholding sustainability. This intersection of sustainability and SDG adherence fosters a framework for hill towns to navigate through their unique challenges, promoting a balanced coexistence between nature and human settlements.

Srinagar is the capital city of Jammu and Kashmir Union Territory in India and is one of the most rapidly urbanizing cities in the Indian Himalayan Region [21]. However, beyond the geographical setting and serenity of the city there is a complicated urban form with narrow streets and densely built neighborhoods. Such a trend has created a challenge to balance urban growth and sustainable development of the city. In this context, there are limited studies which attempt to focus on a comprehensive understanding of the spatiotemporal urban expansion in Srinagar Valley over time. The literature has not adequately addressed the mapping and profiling of Srinagar city using geoinformatics. There is a need for studies that explore the relationship between urban governance, policy changes, and settlement patterns. Identifying and proposing sustainable development strategies by the unique characteristics and challenges of Srinagar Valley is required. Such studies based on historical analysis would comprehensively capture the trajectory of urban development in Srinagar and project future urbanization trends.

In this context, this study investigates the urban dynamics of settlement patterns in Srinagar city, facilitated by the utilization of Geospatial techniques. The primary objective of this study is to identify time-based variations in these settlement patterns and analyze the impact of urban governance and policy changes.

2 About Study Area

Srinagar is the capital city of Jammu and Kashmir union territory of India (Fig. 1). It is situated in the Kashmir Valley along the banks of the Jhelum River, enveloped by the mighty Himalayan ranges. The city experiences a temperate climate, characterized by mild summers and extreme winters with occasional snowfall. It covers an area of 304.27 km^2, and Shuttle Radar Topography Mission (SRTM) digital elevation data has been used to exhibit a variation in elevation ranging from 1558 to 2813 m above mean sea level. Srinagar UA is the second largest city in the Himalayan region after Kathmandu in Nepal. The economy traditionally hinges on sectors like handicrafts, horticulture, and tourism, the latter significantly fueled by the city's enchanting landscapes including the iconic Dal Lake. The architectural heritage marked by ancient temples, forts, palaces, wooden mosques, Mughal gardens, underpins the historical richness of the city while intertwining with burgeoning modern infrastructures, thus portraying a blend of antiquity and contemporary advancement.

2.1 Demographics

The geographic extent of Srinagar UA is sprawling across three districts of Jammu and Kashmir, namely Srinagar, Badgam and Pulwama. Demographically, Srinagar UA constitutes of three statutory towns (ST) of Srinagar Municipal Corporation, Badami Bagh cantonment board and Pampore Municipal Committee; along one census towns (CT) of *Kral Pora* and seven Out Growth (OG) areas. According to the 2011 census, the population of Srinagar Urban Agglomeration (UA) was 1.264 million with a population density of approximately 703 individuals per square kilometer [9]. The population growth rate of Srinagar UA between 2001 and 2011 was 27.93% (Table 1). Post independence, the rapid population growth can be attributed to the migration to the urban areas, increased birth rates and subsequent fall in death rates as well. However, the growth has been scattered due to the geographical features and limitation of the region. The population density does not show any significant change in the Srinagar Urban Agglomeration since the administrative boundaries were subsequently revised through the merger of 62 villages in the 1971 and then later more rural areas were added in 2001.

As Srinagar Master Plan—2031 [22] states "*The sluggish growth rate of Srinagar city is an obvious indicator of its declining economic growth and its creeping into metropolitan stagnation and regional suburbanization.*"

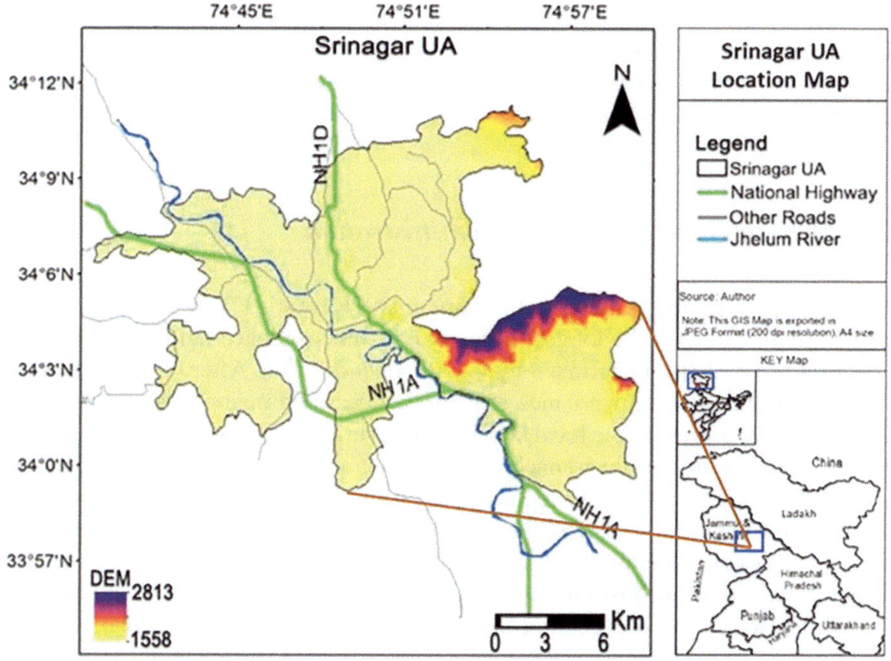

Fig. 1 Study area map (*Source* Author)

Table 1 Growth pattern observed in Srinagar UA

Census year	Area in square kilometer	Persons	Variation since the preceding census		Males	Females
			Absolute	Percentage		
1901		122,618	—	—	65,542	57,076
1911		129,692	7,074	5.77	70,229	59,463
1921		145,229	15,537	11.98	78,545	66,684
1931		178,328	33,099	22.79	97,632	80,696
1941		214,042	35,714	20.03	116,432	97,610
1951		255,731	41,689	19.48	138,066	117,665
1961	50.96	297,421	41,690	16.30	159,700	137,721
1971	107.17	431,835	134,414	45.19	232,786	199,049
1981	210.65	616,835	185,000	42.84	329,737	287,098
1991	–	–	–	–	–	–
2001	242.60	988,210	–	–	537,512	450,698
2011	304.27	1,264,202	275,992	27.93	667,203	596,999

With this information regarding demographic patterns and urban growth challenges, the study further delves into the urban governance structure of the city and understands the relationship and dependency between urban sprawl and changes in governance structure over a period of time.

2.2 *Urban Governance and Administration*

Srinagar being summer capital of the Jammu and Kashmir UT, was part of Jammu and Kashmir State for the longest duration post-independence before the delimitation exercise of 2019, undertaken by Government of India. After the abrogation of Article 370, the urban governance structure witnessed a major overhaul that gave administrative powers to the local bodies at both rural and urban scales. This process led to a change in the governance and administrative structure of the state, with the local-level governance tier coming into being through the introduction of the Panchayati Raj system at the rural level and municipalities at the urban level [23].

As witnessed in many cities in India, the urban development and governance structure have multiple institutions functions in close coordination with one another. The Srinagar city is no exception to this phenomenon where the Urban Planning Authority, i.e., the Srinagar Development Authority (SDA) mandates the preparation and implementation of Master Plan documents, urban local bodies like Municipal Corporations [24] and Municipal Committees govern the issues related to property taxation, water supply, drainage, and sewerage disposal, public street maintenance, building regulations, sanitation and public health, fire safety, public marketplaces, and other public welfare-related issues, keeping in consideration the demographics of the urban government.

3 Methodology and Dataset

The data on urban growth from 1930 to 1980 was sourced from the Town Planning Organization Kashmir [22]. Multi-temporal Landsat satellite images for the study area were selected for the years 1991, 2001, 2011, and 2021 to identify time-based variations in settlement patterns (Table 2). Satellite images and development plans were strategically included in this study to identify chronological variations and enhance understanding of the most recent governance profiles or urban development policies. The methodology adopted in this study is illustrated in Fig. 2.

This study includes the analysis of governance profiles (Urban Development Policy) and settlement patterns (Urban Morphology) within the Srinagar Urban Agglomeration (UA). The urban morphology of the study area was analyzed through Landsat satellite images obtained from the open-source USGS Earth Explorer website (https://earthexplorer.usgs.gov), ensuring images with minimal cloud cover and snow (<10%) of the same season in the late month of April for the maximum accuracy. The

Table 2 Detailed description of Landsat imagery metadata

S. no	Satellite/sensors	Acquisition date	Pixel size	Path/row
1	Landsat-5/TM	20/04/1991	30 m	149/36
2	Landsat-5/TM	10/04/2001	30 m	149/36
3	Landsat-5/TM	27/04/2011	30 m	149/36
4	Landsat-8/TIRS	30/04/2021	30 m	149/36

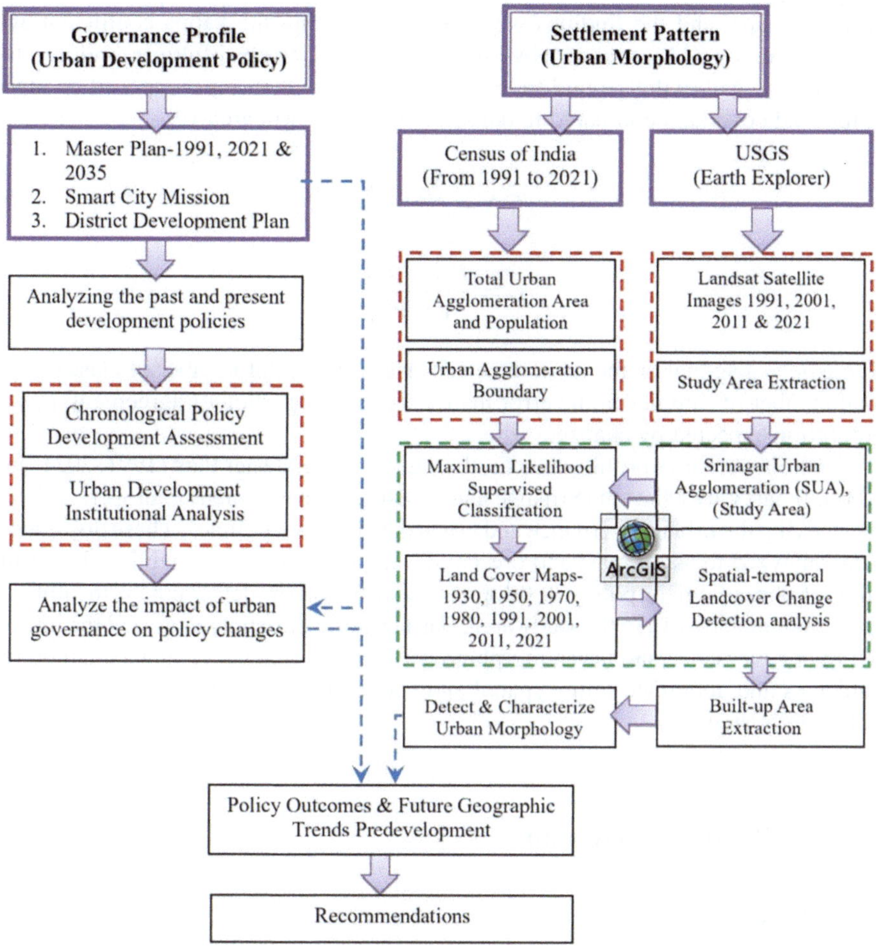

Fig. 2 Flowchart of the methodology adopted in this study (*Source* Author)

detailed description of the selected satellite imagery metadata is shown in Table 1. Data on the total urban agglomeration area, population, and boundary of the study area were sourced from the census of India website (https://censusindia.gov.in/census.website/data/population-finder).

The maximum likelihood classification tool in ArcGIS 10.3 was utilized to generate spatiotemporal land cover maps, followed by land cover change detection. Subsequently, the built-up area was extracted to identify and characterize urban morphology. The ground truth data was used to verify the accuracy assessment of land use maps, and the findings were interpreted using the kappa coefficient and overall accuracy (OA), as shown by Eqs. 1 and 2, respectively. Multiple random sites were used to gather the ground truth data, which were then superimposed on maps of the land cover and compared to data from Google Earth archives.

$$\mathrm{OA} = \frac{\sum_{i=1}^{k} n_i}{n} \tag{1}$$

$$\kappa_i = \frac{P_{(o)} - P_{(e)}}{1 - P_{(e)}} \tag{2}$$

where n_i = diagonal elements in the error matrix, k = total number of classes, n = total number of samples in the error matrix, $P_{(\mathrm{o})}$ = proportion of agreement, $P_{(\mathrm{e})}$ = proportion expected by chance.

For the governance profile analysis, the data from the Master Plan (1991, 2021, and 2035), Smart City Mission (Srinagar), and District Development Plan were utilized for studying the urban governance and urban development policies. These documents were analyzed to assess past and present development policies, enabling the evaluation of chronological policy development and the urban development institutions (government network) in the study area. Furthermore, this study analyzed the impact of urban governance and urban morphology on policy changes to understand policy outcomes and anticipate future spatial and geographic trends in the redevelopment of the study area.

4 Results and Discussion

4.1 Master Plans

The timeline of the development policy and outcomes of Master Plans for the study area is depicted in Fig. 3. After 1947, Srinagar experienced rapid, unplanned development with very few planned housing complexes. The first-ever comprehensive Master Plan (1971–1991) aimed to ensure sustainable growth after the devastating floods of 1902, which had caused Srinagar to expand mostly in low-lying areas, wetlands, and flood absorption basins near the Flood Spill Channel in the west and southwest. Political turmoil in 1989 led to a ten-year planning hiatus, and weak laws and regulations of urban local bodies and authorities triggered the widespread transformation of residential areas and colonies into commercial developments.

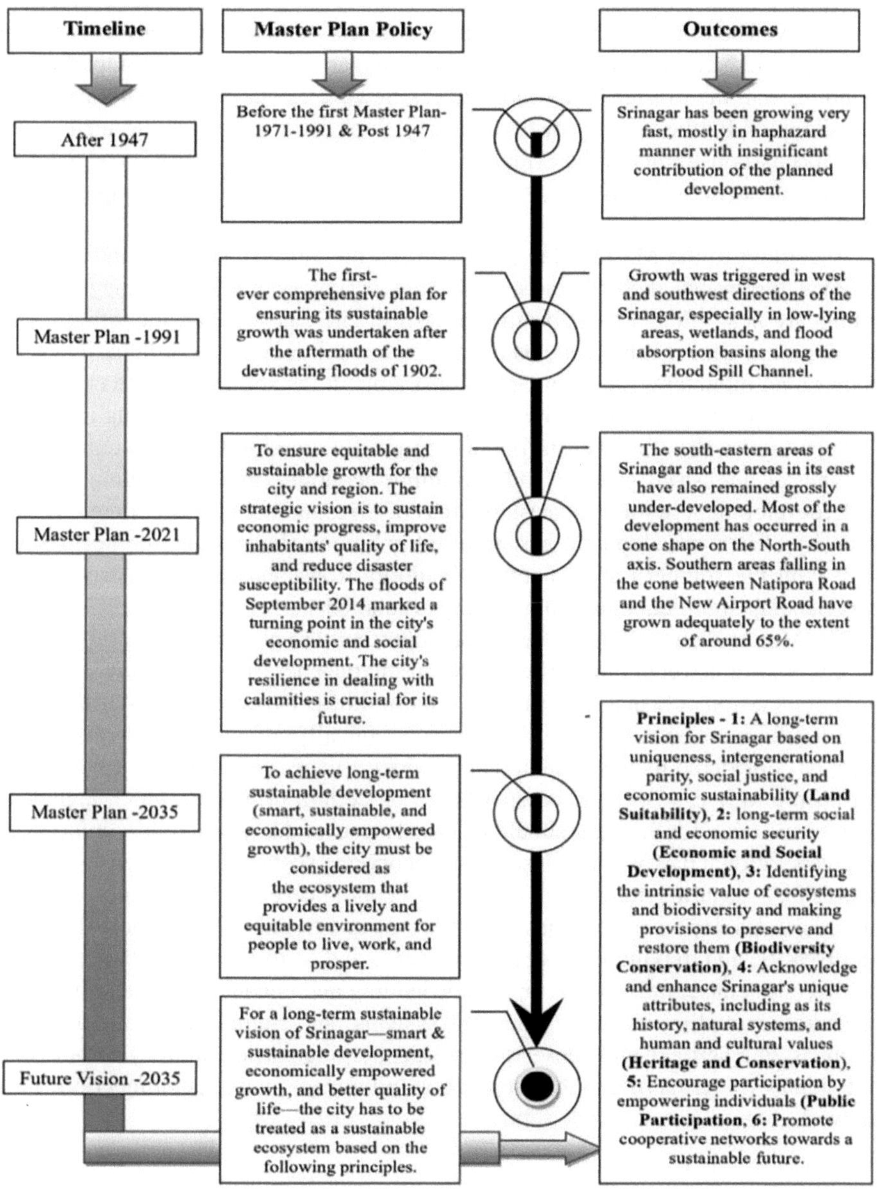

Fig. 3 The timeline framework of development policy and outcomes (*Source* Author)

The process of promoting land use conversions accelerated with the approval of the Srinagar Master Plan (2001–2021), aimed at ensuring equitable and sustainable growth for the city and region. The strategic vision is to achieve sustainable economic progress, improve quality of life, and reduce susceptibility to disasters. The floods of September 2014 marked a turning point in the economic and social development of the city. The city's resilience in dealing with calamities is crucial for its future. Although the Master Plan (1971–1991) of Srinagar offered a more comprehensive planning than the Master Plan (2001–2021); however, both plans did not provide a safe direction for the future growth of the city. Such a scenario might be attributed to the capacity constraints of the Srinagar Municipal Corporation (SMC) and Srinagar Development Authority (SDA), as well as departments responsible for implementing master plans.

According to the Master Plan (2001–2021), the city has been expanding along national highways, district roads, and main periphery roads in Srinagar, albeit cautiously near silted flood absorption basins. The Master Plan identified lateral development trends, constraints on expansion toward flood absorption basins, water bodies, and mountains, seismography constraints in the southwest. The constraints on vertical development due to poor soil-bearing capacity, deteriorating arboriculture conditions, rising land costs, population growth, immigration and outmigration, imbalanced industrial growth, and dwindling economic growth, leading to gross underdevelopment in the southeastern areas of Srinagar and its eastern regions. The majority of new development has occurred in a cone shape along the north–south axis, with the southern areas between Natipora Road and the New Airport Road experiencing significant growth, reaching around 65%.

The Master Plan 2035 is viewed as an opportunity to plan for transformation and the city's sustained economic growth. The master plan describes the future strategy in light of its vision of Srinagar as a city of opportunities. To achieve long-term sustainable development (smart, sustainable, and economically empowered growth), the city must be considered the ecosystem that provides a lively and equitable environment for people to live, work, and prosper based on certain principles which includes a long-term vision for Srinagar should be based on several key principles: uniqueness, intergenerational parity, social justice, and economic sustainability (focused on land suitability); ensuring long-term social and economic security (through economic and social development strategies); identifying the intrinsic value of ecosystems and biodiversity, and making provisions to preserve and restore them (in the context of biodiversity conservation efforts); acknowledging and enhancing Srinagar's unique attributes, including its history, natural systems, and human and cultural values (emphasizing heritage and conservation); encouraging participation by empowering individuals (in terms of public participation initiatives); and promoting cooperative networks toward a sustainable future.

4.2 Governance Network

Until the national government's delimitation process in 2019, Srinagar city, served as the summer capital of the Jammu and Kashmir State since the independence of India in 1947. As a result of this process, the administrative and governing structure of the state changed, and with the repeal of Article 370, the Panchayati Raj system was introduced at the rural level, establishing the local level governance layer. Figure 4 shows the adopted network of governing bodies in the study area. The urban development and governance network system in Srinagar is characterized by several entities like the Housing and Urban Development Department (HUDD), Rural Department (RD), Badami Bagh Cantonment Board (BBCB), Irrigation and Food Control (I&FC) Department, Power Development Department (PDD), Public Works Department (PWD), Lakes Development Authority (LDA), Higher Education Department-J&K (JKHE), and Srinagar Smart City Limited (SSCL) that work closely with one another. Srinagar Development Authority (SDA) coordinates with all the parastatal agencies for the preparation and implementation of Srinagar Master Plan.

The decimation of authority for the urban governance is directly from the central level agencies since the delimitation and formation of Jammu and Kashmir Union Territory. Such correlations often address to quick decision-making scenarios and

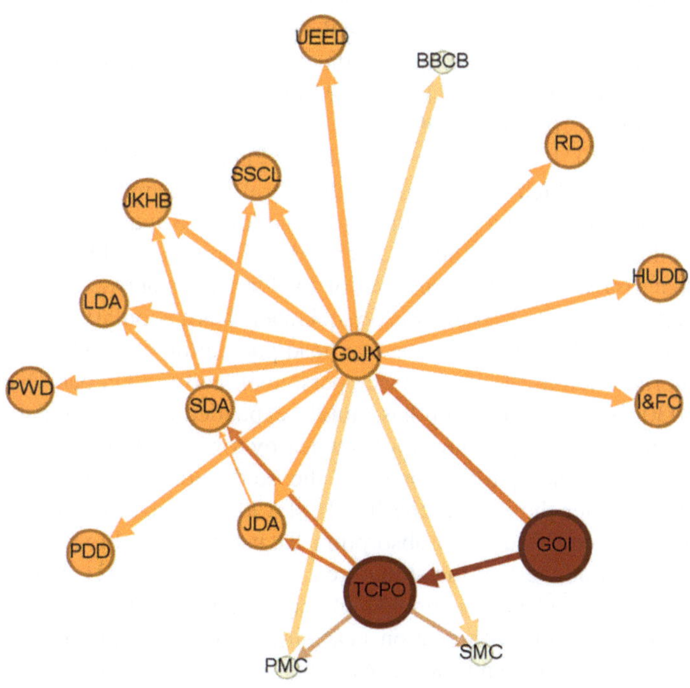

Fig. 4 Adopted network of governing body (*Source* Author)

enhancement in the political willingness within the institutions, but is often critically accessed as non-federal. In the past there have been demand for complete state-hood thus delegating more governance power to the local and state level agencies for UTs like Delhi, Chandigarh, etc. The urban governance and institutional network will again witness major changes in the structure and dependencies once the state-hood of Jammu and Kashmir UT is restored.

Although the growth patterns and corridors developed within a geography are an outcome of many factors including, geographical limitations, critical infrastructure investments in the region, political background, local aspirations, etc., and to realize these changes in the physical form surely is a time taking process.

4.3 Growth Patterns and Growth Corridors

The overall accuracy (OA) and kappa coefficient (k) metrics were obtained for the land cover maps of 1991 to 2021.The result indicates a strong agreement between ground truth data and categorized maps. The overall accuracy of the land cover maps ranged from 85 to 95% and kappa coefficient was between 0.80 to 0.82. Therefore, a minimum 80% accuracy threshold was met, indicating that the classification procedures were successful. Figure 5 shows the patterns of historical built-up growth in Srinagar Urban Agglomeration from 1930 to 2021. It can be observed that there was slow built-up growth in the city during the post-independence period after 1947, followed by rapid built-up growth after the 1970s. The total built-up area within the study area increased from 12 km^2 in 1930 to 19 km^2 in 1950, 24 km^2 in 1970, 30 km^2 in 1980, 39.70 km^2 in 1991, 55.73 km^2 in 2001, and 62.75 km^2 in 2011, and 85.46 km^2 in 2021. According to the Master Plan of 2035, the Greater Srinagar Metropolitan limit, which was notified as the Local Area Limits of SDA, expanded from 416 to 766 km^2 to include the municipal areas of the Srinagar Municipal Corporation as well as the ULBs of Budgam, Ganderbal, Pampore, and Khrew, along with an additional 160 villages that grew out of twelve tehsils across six districts. The Srinagar Metropolitan Region covers 766 sq. km and has a population of 0.17 million (Census of India 2011), excluding the overhead population of roughly 0.03 million, which comprises defense forces, service personnel, and the Darbar. Over the past century (1930–2021), the built-up has rapidly expanded due to population growth and migration for better opportunities from nearby towns and villages.

The analysis of the spatial growth of the study area reveals an exponential growth pattern. After the devastating floods of 1902, Srinagar expanded into low-lying areas, wetlands, and flood absorption basins near the Flood Spill Channel in the west and southwest of the study area. From 2001 to 2021, built-up growth predominantly occurred along national highways, district roads, and main periphery roads in Srinagar, albeit with caution near silted flood absorption basins. Identified lateral development trends and constraints on growth toward flood absorption basins, water bodies, and mountains, as well as seismography constraints in the southwest, limitations on vertical development due to poor soil-bearing capacity,

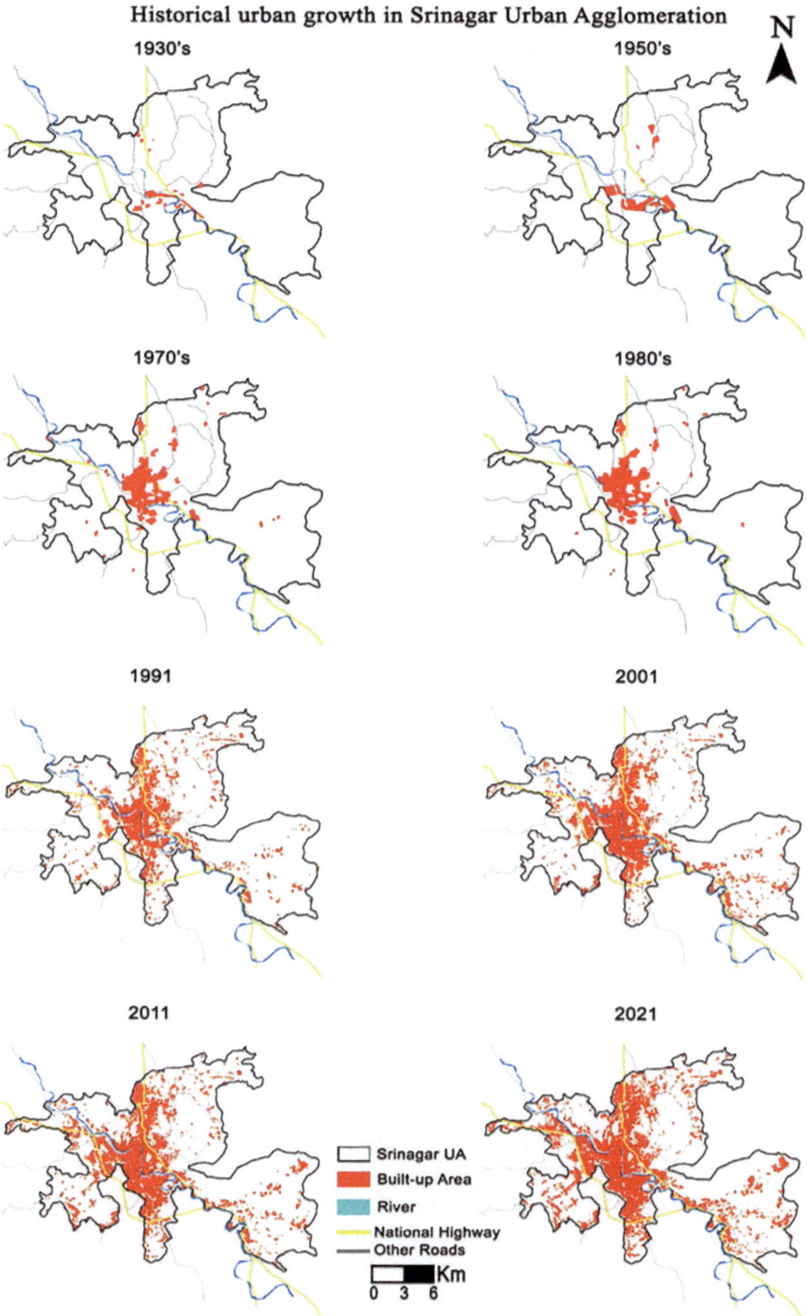

Fig. 5 Historical urban growth in Srinagar urban agglomeration (*Source* Author)

deteriorating arboriculture conditions, rising land costs, population growth, immigration and outmigration, imbalanced industrial growth, and dwindling economic growth, have contributed to the gross underdevelopment of the southeastern areas of Srinagar and its eastern regions. Most built-up development has taken the shape of a cone along the north–south axis. Southern areas within this cone, situated between Natipora Road and the New Airport Road, have seen substantial growth.

4.4 SDG's Focused Smart City Initiatives

Srinagar Smart City has set an inspiring standard with visionary initiatives such as "Srinagar Square," the restoration of ghats to rejuvenate water transport, the revitalization of the breathtaking lakefront, the preservation of the city's rich heritage, and the beautification of underutilized spaces. These initiatives have made Srinagar a shining example of innovation for urban development professionals.

The development of Srinagar Smart City has initiated people-centric approaches toward the SDGs, incorporating elements of traditional art and culture, heritage preservation, smart facilities, efficient governance, and sustainability, all to ensure social fairness and improve residents' quality of life.

5 Discussion and Conclusion

The historical trajectory of urban development in Srinagar reveals a complex interplay of factors, from rapid post-independence growth to subsequent planning challenges due to political turmoil. Initial efforts, like the Master Plan 1971–1991, aimed to mitigate the impacts of natural disasters and guide sustainable development. However, a lack of regulatory enforcement during periods of instability led to unchecked urban sprawl and the conversion of residential areas into commercial zones. Despite subsequent planning endeavors, such as the Master Plan 2001–2021, the city continues to grapple with the consequences of inadequate infrastructure and uneven development, highlighting the need for more effective governance mechanisms.

Further, analyzing growth patterns and constraints provides insights into the spatial dynamics of urban expansion in Srinagar. The city witnessed slow growth post-independence, followed by rapid urbanization concentrated along transportation corridors. However, development has been constrained by factors such as flood absorption basins, topographical features, and socio-economic disparities. Notably, the underdevelopment of southeastern areas underscores the importance of adopting inclusive and balanced growth strategies to address spatial inequalities and promote sustainable urbanization.

The governance framework for urban development in Srinagar is characterized by a complex network of departments and authorities operating at different levels. Recent shifts in governance structures, including the transition from state to local administration, have reshaped administrative dynamics. Collaborative efforts among entities like the Housing and Urban Development Department, Srinagar Development Authority, and Smart City initiatives signify a concerted approach toward addressing urban challenges. The emergence of Srinagar Smart City represents a significant milestone in promoting sustainable urban development. Through visionary initiatives like "Srinagar Square" and heritage preservation projects, the Smart City initiative aims to integrate traditional art, culture, and sustainability principles into urban planning. By fostering socio-economic inclusivity and enhancing residents' quality of life, these initiatives align with the broader objectives of the Sustainable Development Goals (SDGs). Moving forward, continued investment in smart infrastructure and community-driven initiatives will be essential for realizing the full potential of Srinagar as a model for innovative and people-centric urban development.

Overall, the analysis of historical urban governance in Srinagar reveals a strong influence on settlement patterns. During these years, shifts in political and administrative structures have led to distinct changes in the growth pattern of the city. Centralized governance has historically resulted in concentrated development in specific areas, contributing to uneven settlement patterns of urban expansion along the north–south axis, with certain areas experiencing rapid urbanization while others remain underdeveloped in the southeast and eastern areas of Srinagar. After the abrogation of Article 370, placing the city under the Union Territory (UT) changed the political landscape, necessitating a focus and adjustment in governance networks and social infrastructure to align with experiences of other Indian cities.

The settlement pattern of Srinagar highlights the importance of considering historical, geographical, economic, and policy factors in urban planning and development. As the city evolves, policymakers and stakeholders must adopt sustainable strategies that balance growth with environmental conservation, cultural preservation, and equitable access to resources and opportunities. Addressing the complex challenges of urban development in Srinagar requires a multifaceted approach that encompasses effective governance, inclusive growth strategies, and innovative urban planning initiatives. By embracing principles of sustainability, equity, and resilience, Srinagar can chart a path toward a more vibrant, inclusive, and resilient urban future.

References

1. Tiwari P, Tiwari A, Joshi B (2018) Urban growth in Himalaya: understanding the process and options for sustainable development. J Urban Reg Stud Contemp India 4(2):15–27
2. Liu C, Cheng L, Li J, Lu X, Xu Y, Yang Q (2023) Trade-offs analysis of land use functions in a hilly-mountainous city of northwest Hubei Province: the interactive effects of urbanization and ecological construction. Habitat Int 131:102705. https://doi.org/10.1016/j.habitatint.2022.102705
3. Schickhoff U, Singh RB, Mal S (eds) (2022) Mountain landscapes in transition effects of land use and climate change. Springer. https://doi.org/10.1007/978-3-030-70238-0
4. Anker Y et al (2019) Effect of rapid urbanization on Mediterranean karstic mountainous drainage basins. Sustain Cities Soc 51(June):101704. https://doi.org/10.1016/j.scs.2019.101704

5. Kumar A, Pushplata (2015) City profile: Shimla. Cities 49:149–158. https://doi.org/10.1016/j.cities.2015.08.006
6. Pradhan Q (2007) Empire in the hills: the making of hill stations in Colonial India 23(1). https://doi.org/10.1177/025764300602300102
7. Badar RN, Bahadure SP (2020) Assessing tourism sustainability in hill towns: case study of Shimla, India. J Mt Sci 17(9):2241–2261. https://doi.org/10.1007/s11629-019-5683-5
8. Bhardwaj A, Gupta AK (2023) Tourism sustainability in hilly regions—a review for Shimla. In: Hau KK, Gupta AK, Chaudhary S, Gupta T (eds) Recent advances in structural engineering and construction management, pp 873–881. Springer Nature, Singapore
9. Chettry V, Surawar M (2021) Assessment of urban sprawl characteristics in Indian cities using remote sensing: case studies of Patna, Ranchi, and Srinagar. Environ Dev Sustain 23(8):11913–11935. https://doi.org/10.1007/s10668-020-01149-3
10. Chandel P, Anand S, Singh D (2022) An overview of scientific research on geoheritage in India. Geoheritage 14:131. https://doi.org/10.1007/s12371-022-00762-0
11. Mir AR, Dar FA, Ahmad MZ (2023) Characteristics of geosites for promotion and development of geotourism in Ladakh, India. Geoheritage 15(3):1–16. https://doi.org/10.1007/s12371-023-00866-1
12. Bharti N, Khandekar N, Sengupta P, Bhadwal S, Kochhar I (2020) Dynamics of urban water supply management of two Himalayan towns in India. Water Policy 22:65–89. https://doi.org/10.2166/wp.2019.203
13. Negi VS, Tiwari DC, Singh L, Thakur S, Bhatt ID (2022) Review and synthesis of climate change studies in the Himalayan region. Environ Dev Sustain 24(9):10471–10502. https://doi.org/10.1007/s10668-021-01880-5
14. Upadhaya K, Pandey HN, Law PS, Tripathi RS (2003) Vascular plant diversity in the sacred groves of Jaintia Hills in northeast India. Biodivers Conserv 12(7):1497–1510. https://doi.org/10.1023/A:1023682228549
15. Joshi N (2021) Adopting a governance lens to address urban risks in the Uttarakhand Himalayas: the case of Almora, India. Int J Disaster Risk Reduct 54:102044. https://doi.org/10.1016/j.ijdrr.2021.102044
16. Chettry V (2023) Geospatial analysis of urban sprawl in Agartala municipal council, India, from 1991 to 2021. In: Filho WL, Ng TF, Iyer-Raniga U, Ng A, Sharifi A (eds) SDGs in the Asia and Pacific region, implementing the UN sustainable development goals—regional perspectives, pp 1–25. Springer Nature, Switzerland. https://doi.org/10.1007/978-3-030-91262-8_112-1
17. Chettry V, Manisha K (2022) Assessing and predicting urban growth patterns using ANN-MLP and CA model in Jammu urban agglomeration, India. In: Das B, Patgiri R, Bandyopadhyay S, Balas VE (eds) Modeling, simulation and optimization. Smart innovation, systems and technologies, pp 387–397. Springer. https://doi.org/10.1007/978-981-19-0836-1_30
18. Sen A, Choudhury D, Chettry V (2023) Investigation of bioclimatic design features in vernacular architecture of Northeast India: case studies of Assam-type houses and stilt houses (Chang Ghar) in Assam, India. Indian J Tradit Knowl 22(03):659–668. https://doi.org/10.56042/ijtk.v22i3.5757
19. Dasgupta R et al (2023) Exploring indigenous and local knowledge and practices (ILKPs) in traditional Jhum cultivation for localizing sustainable development goals (SDGs): a case study from Zunheboto district of Nagaland, India. Environ Manage 72(1):147–159. https://doi.org/10.1007/s00267-021-01514-6
20. Chhetri R, Kumar P, Pandey VP, Singh R, Pandey S (2020) Vulnerability assessment of water resources in Hilly region of Nepal. Sustain Water Resour Manag 6(3):1–12. https://doi.org/10.1007/s40899-020-00391-x
21. Parry JA, Ganaie SA, Sultan Bhat M (2019) GIS based land suitability analysis using AHP model for urban services planning in Srinagar and Jammu urban centers of J&K, India. J Urban Manag 7(2):46–56. https://doi.org/10.1016/j.jum.2018.05.002
22. Town Planning Organisation, Kashmir (2019) Master plan 2035, Srinagar metropolitan region
23. Impact of abrogation of article 370 on J&K. https://pib.gov.in/Pressreleaseshare.aspx?PRID=1694784. Accessed 12 Nov 2023

24. Srinagar municipal corporation Act 2000 (2000) https://www.indiacode.nic.in/bitstream/123456789/9595/1/municipal_corporation_act%2C_2000.pdf. Accessed 22 Nov 2023

Satellite Remote Sensing Techniques for Air Quality Monitoring in the Mid-Brahmaputra Valley of Northeast India

Nivedita Barman

1 Introduction

Aerosols, which are defined as solid or liquid particles suspended by a gas or combination of gases [1, 2], continue to be a significant source of uncertainty in estimates of climate forcing [2, 3]. The word "aerosols" refers to a vast category of minute particles, including sulfuric acid droplets, pollen, mineral dust, sea salt particles, and fog, among many others. The simplest yet evident example of atmospheric aerosol is the cloud which is a suspension of condensed water droplets in air. There are many different natural or man-made sources of atmospheric particles. Natural sources of emissions include soils, vegetation, fires, volcanoes, and oceans. However, anthropogenic particle sources include emissions from burning fossil fuels, biofuels, other fuels (including peat), and human-caused vegetation fires [3, 4]. Different processes involved in the formation of particles are Bulk-to-Particle Conversion (BPC), Gas-To-Particle Conversation (GPC), and combustion [2, 5]. In a broad spectrum, particles can be classified as primary or secondary based on the source of origin. Figure 1 summarizes different types of aerosols. Atmospheric aerosols may be primary, which are directly emitted as a particle into the atmosphere, or secondary, which are produced from gases through gas-to-particle conversion processes [6]. They have typical radii ranging from 0.001 to 100 μm and of various shapes. Based on their size; aerosols are classified into nucleation, accumulation, and coarse mode particles [7]. Particles with sizes less than 50 nm diameter are commonly referred to as the nucleation mode. Aerosol particles in the size range 50 nm–1 μm in diameter are accumulation mode particles and particles greater than ~1 μm in diameter are coarse mode. After emission, particles undergo various physical and chemical interactions and transformations in terms of particle size, structure, and composition [1]. Particles can be removed from the atmosphere in a variety of methods, depending

N. Barman (✉)
Department of Environmental Science, Tezpur University, Tezpur, India
e-mail: niveditabarman3@gmail.com

V. Chettry et al. (eds.), *Geoinformatics for Hilly Regions*, Advances in 21st Century Human Settlements, https://doi.org/10.1007/978-981-95-0156-4_8

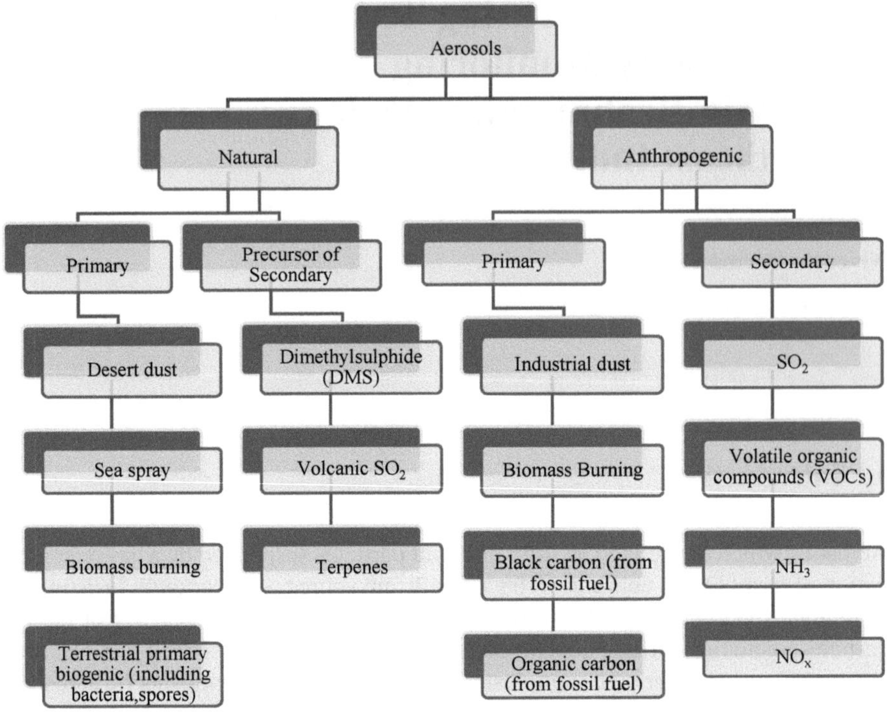

Fig. 1 Aerosol types [4] (*Source* Author)

on their size and properties. A yearly worldwide mean indicates that in-cloud and below-cloud scavenging (wet deposition) removes 80–90% of the particles from the atmosphere, with dry deposition—which includes impaction of particles on different surfaces—removing the remaining particles [3].

Atmospheric particles have a large impact on climate and health. Because of their significant impact on the global radiative balance, particles may be the primary cause of climate change. Two avenues exist for particles to impact the climate. The scattering of solar radiation and the absorption/emission of terrestrial radiation have a direct impact on Earth's radiation balance. The indirect effect on radiative forcing is via their role in cloud formation as particles can modify cloud properties [8–11]. Particles also affect vegetation, ice, and snow [4]. Fine particles (nucleation and accumulation mode) pose severe threat to human health by causing several cardiovascular, respiratory, and allergic diseases as they are able to get past the main airway defenses and reach the furthest reaches of the lungs, known as alveoli [1, 11]. Coarse particles can irritate a person's eyes, nose, and throat. The term particulate matter (PM) commonly encompasses the solid phase of suspended matter in the atmosphere. Of the six criteria pollutants, particulate matter is the most significant in terms of its detrimental impacts on human health [12]. Fine particles have longer lifetimes and can travel long distances while larger coarse particles settle faster and are not readily transported over longer distances [12–14]. The main sources of PM_1, $PM_{2.5}$, and PM_{10} were found to include biomass burning, wind-blown dust, secondary aerosol, coal combustion, vehicle exhaust, etc. [15].

Black carbon (BC) is the sooty black material released during the combustion of fossil fuels or biomass [16–18]. Burning biomass fuels for cooking, such as wood, manure, and crop residue, is the main cause of indoor BC. When burning fossil fuels like coal and diesel outside, burning open biomass that is linked to deforestation and burning agricultural residue, and using biofuels for cooking are the main outdoor sources of BC emissions [17]. Because BC is a powerful absorber of solar radiation, it might have a major regional impact on precipitation, cloud cover, and temperature [17, 19–21].

These days, air pollution is one of the biggest issues. Because atmospheric particles are one of the main contributors to air pollution, research on them is becoming more and more important globally. It is easier to implement effective control programs when trends and variability in the concentration of particles in ambient air are studied [22]. However, there isn't much research on aerosols from north-eastern India. The present study examines columnar and surface loading of aerosols over north-eastern parts of Indian land mass in different seasons using remote sensing observations from satellites and ground-based in-situ measurements by focusing on the following objectives:

1. Temporal variability of surface and columnar aerosol concentration in the Brahmaputra Valley in Assam and their comparison.
2. Establish a relationship among PM_{10}, $PM_{2.5}$, and BC concentration of the Brahmaputra Valley in Assam.

2 Study Area and Meteorology

The study site is the Tezpur University campus (26.70 °N and 92.83 °E, elevated about 48 m above mean sea level) which is located in Napaam, a suburb, about 15 km east of Tezpur city. There are villages and verdant landscapes all around the site. Tezpur is an ancient city on the banks of the river Brahmaputra and is the second largest of the north bank towns with a population of 100,477 (according to the 2011 census). The star in Fig. 2 shows the study location in the north-eastern part of India.

The city falls in the subtropical climatic zone and experiences moderate to high rainfall [23]. High humidity exists in the area, especially during the period, from May to November. As the city is situated amidst several small hillocks, flooding during the monsoons is usually not seen. The Indian Meteorological Department (IMD) recognizes four distinct seasons, which are taken into consideration in this study to analyze the seasonal fluctuations of aerosols: pre-monsoon (March–May), monsoon (June–September), post-monsoon (October–November), and winter (December-February). Mostly calm condition prevails in the area except for the Bordoichila or Nor'westers during March–April (a wild wind in Assam with thunderstorms that marks the incoming of monsoon season).

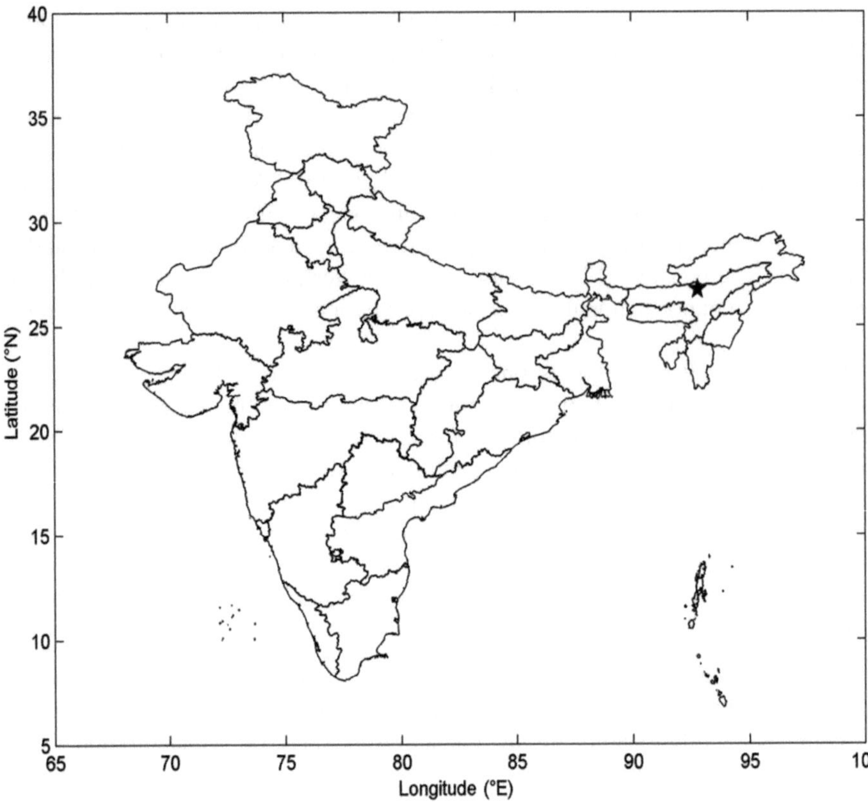

Fig. 2 Location of study region (star) in Northeast India (*Source* Author)

3 Data and Methodology

The methodology adopted in this study is presented in Fig. 3. Using daily Aerosol Optical Depth (AOD) observations from the Moderate Resolution Imaging Spectrometer (MODIS) onboard the Earth studying System (EOS) Terra satellite, which has been studying the Earth since mid-2002, columnar aerosol loading over the study region is analyzed [24]. Seven wavelengths over ocean (470, 550, 660, 870, 1240, 1630, and 2130 nm) and three wavelengths over land (470, 550, and 660 nm) are included in the MODIS products [25]. The present study uses daily AOD at 550 nm retrieved using MODIS Level 3 atmosphere aerosol product data for two years, from July 2013 to June 2015. Around 10:30 am is when the Terra orbit crosses the local equator. NASA's MODIS interactive visualization and analysis tool GIOVANNI (http://giovanni.gsfc.nasa.gov/giovanni) provided daily temporal and spatial resolutions of AOD satellite data for the study area at 550 nm (Dark target) [26].

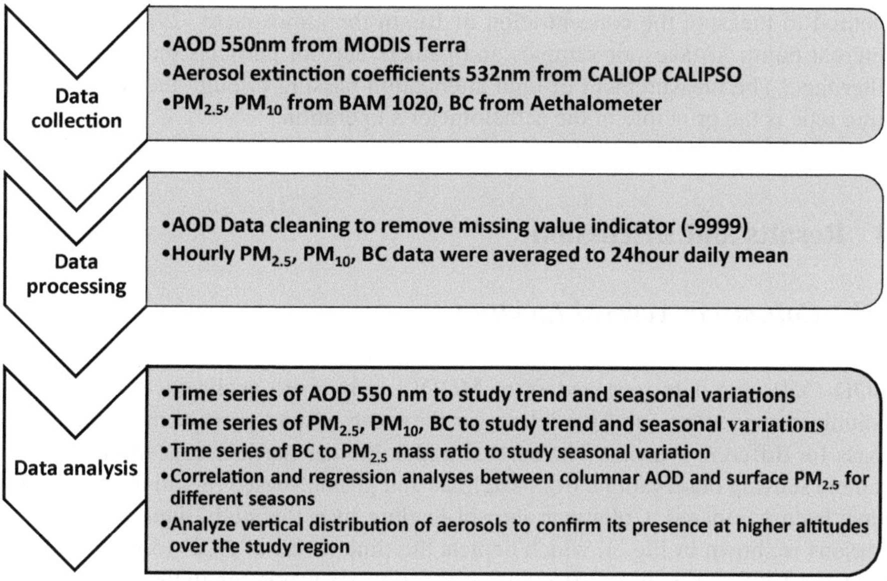

Fig. 3 Flowchart of methodology (*Source* Author)

Using vertical profiles of aerosol extinction coefficients at 532 nm from Cloud Aerosol Lidar with Orthogonal Polarization (CALIOP) onboard Cloud Aerosol Lidar and Infrared Path-finder Satellite Observations (CALIPSO), the vertical distribution of aerosols over the research region is investigated [27].

To examine the surface concentration of aerosols, real-time measurements of $PM_{2.5}$, PM_{10} and BC mass concentrations for the same period of two years, July 2013–June 2015, are used. The Beta Attenuation Monitor (BAM-1020; Met One Instruments, USA) was utilized to quantify the mass concentrations of $PM_{2.5}$ and PM_{10}, utilizing the principle of beta ray attenuation to automatically measure and record the levels of airborne particle concentration in micrograms per cubic meter (μg/m^3). A size-selective inlet, a filter tape, a beta radiation source, and a beta radiation detector make up the automatic BAM. On a filter tape, particles smaller than the size-selective inlet's cutoff diameter are gathered at a single location. A sensitive scintillation detector detects and counts high-energy electrons (beta particles) emitted at a steady rate from a tiny Carbon 14 (C-14) source. An external pump draws a predetermined volume of dust-laden air through a filter tape, which is then automatically positioned between the source and the detector to attenuate the beta particle signal once the tape is completely loaded with dust. The mass concentration of particulate matter on the filter tape and, consequently, the volumetric concentration of particulate matter in the surrounding air are determined by measuring the degree of attenuation of the beta particle signal [28].

A seven-channel Aethalometer (Model AE-16, Magee Scientific USA) was used to measure BC in real-time. Aethalometers use the filter-based optical attenuation method to measure the concentration of BC in the atmosphere [29]. Using a tiny internal pump, it takes air samples and gathers aerosol particles on a quartz fiber filter tape. The measurement of light attenuation passing through the aerosol-filled filter tape is the principle of the aethalometer's operation.

4 Results and Discussion

4.1 Columnar Aerosol Loading

AOD (550 nm) data retrieved from MODIS terra were analyzed to observe the columnar concentration of particles over the north-eastern parts of the Indian landmass for different seasons. The surface and columnar particle concentration using remote sensing observations from satellites and ground-based in-situ measurements have been compared. Columnar aerosol loading over the study region in different seasons is shown in Fig. 4, which depicts the time series of AOD (550 nm) during the period July 2013–June 2015. Seasonal variations were seen in the AOD concentrations. The Columnar AOD values are observed higher during pre-monsoon and lower in post-monsoon season. Pre-monsoon high of AOD was reported from many other places like Kanpur, Varanasi, Doon Valley, and Dibrugarh [26, 30–33]. Long-range aerosol transport is responsible for the high-columnar aerosol loading observed during the pre-monsoon. The pre-monsoon season was marked by strong winds (Nor'westers) that carried dust particles from distant places to the study site. The high-mixing layer height facilitated the mixing up of particles and they are well distributed in the column. There were also many fire episodes related to agriculture biomass burning and forest fires in pre-monsoon. A decrease in aerosol loading due to heavy rainfall during monsoon can also be observed in Fig. 4. Because of the low temperature and calm wind conditions during the post-monsoon, the aerosols are primarily localized near the surface, resulting in minimal columnar aerosol loading. But there were particular days during the post-monsoon season when AOD values were significantly high. These days coincided with the Diwali celebration. Firework activities during Diwali releases a high amount of particulates which build up the aerosol concentration during those festival days. The high-mixing layer height led to a well ventilation effect and subsequently, the AOD values rose during the Diwali days. The chemicals used in crackers like copper, cadmium, zinc, lead, etc., can result in burns or respiratory episodes, and it also fuels a number of inflammatory conditions and infectious disorders. Unexpectedly, the pre-existing condition may get worse and give rise to a new one. Bursting crackers release more heat, carbon dioxide, and other harmful gases into the atmosphere, raising the planet's temperature and contaminating the air, ultimately resulting in global warming.

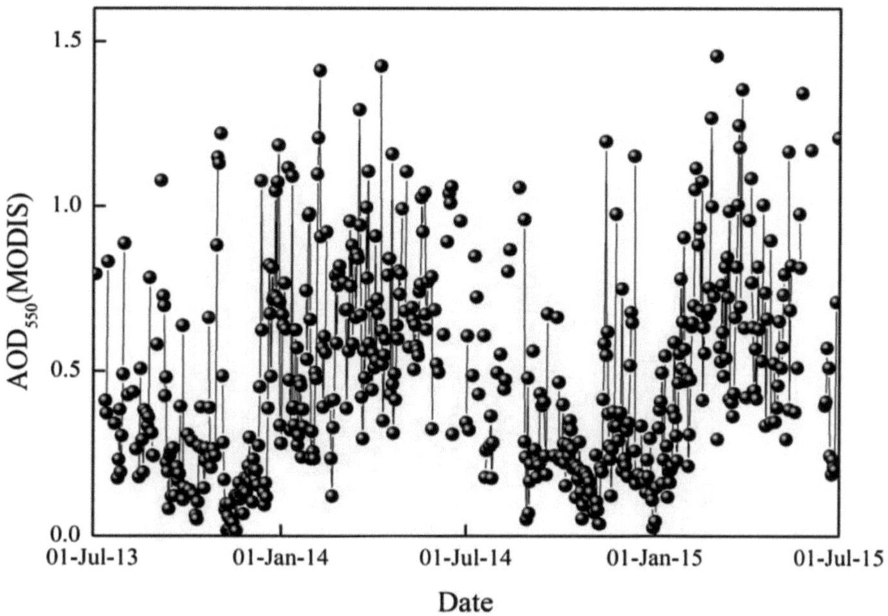

Fig. 4 Columnar aerosol optical depth (550 nm) from MODIS over the study region during July 2013–June 2015 (*Source* Author)

4.2 Surface Concentration of Aerosols

Figure 5 shows daily averaged $PM_{2.5}$, PM_{10}, and BC mass concentrations for two years during July 2013–June 2015, as represented by square, star, and inverted triangle curves, respectively. Open markers represent daily mean values and the vertical bars represent corresponding standard errors. The highest and lowest mass concentrations of $PM_{2.5}$, PM_{10}, and BC were observed in December and July, respectively. Clear seasonality was seen in the case of BC, PM_{10}, and $PM_{2.5}$ mass concentrations. The winter season showed the highest levels of BC, PM_{10}, and $PM_{2.5}$, followed by the pre-monsoon and post-monsoon seasons. Such seasonal variation was also observed at other locations like Bhubaneswar [34, 35]. Winter maximum can be attributed to anthropogenic biomass burnings, low temperature, and other meteorological conditions [34–36]. The levels of BC, PM_{10}, and $PM_{2.5}$ were lowest in the monsoon season due to rain scavenging.

To examine the fractional contribution of BC in the total aerosol system, the ratio of BC to $PM_{2.5}$ is estimated as shown in Fig. 6, which depicts the time series of BC to $PM_{2.5}$ mass ratio from July 2013–June 2015. The seasonal fluctuations in the ratio of BC to $PM_{2.5}$ are seen as completely distinct from the noted shifts in the mass concentrations of $PM_{2.5}$, PM_{10}, and BC. Though BC concentration was highest during winter, its fractional contribution is found highest in monsoon and lowest in pre-monsoon followed by winter. This could be due to the difference in efficiency of

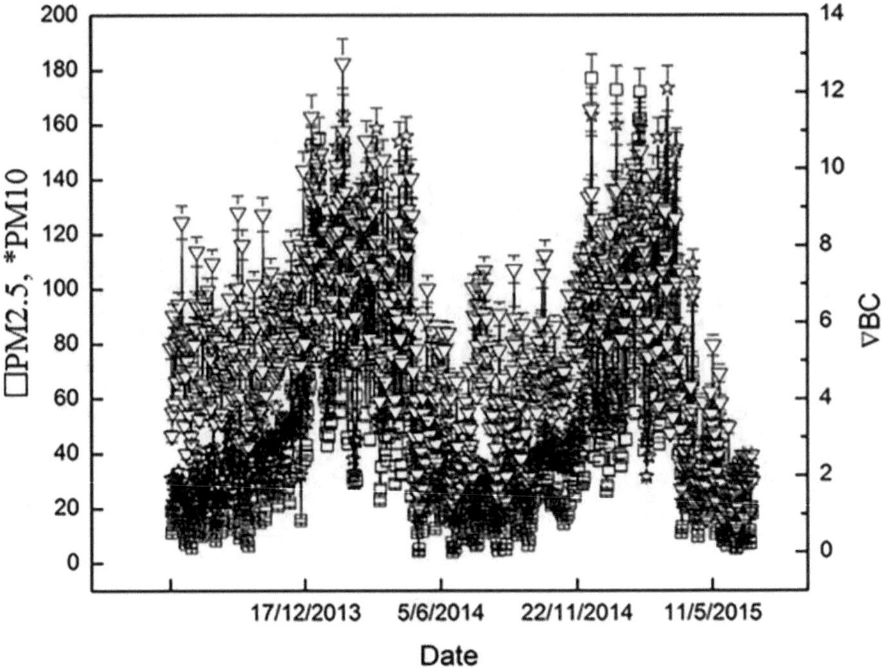

Fig. 5 Daily mean mass concentrations of $PM_{2.5}$ (square), PM_{10} (star), and BC (inverted triangle) over Tezpur during July 2013–June 2015. Open markers represent daily mean values and the vertical bars represent corresponding standard errors (*Source* Author)

the wet removal processes on black carbon compared to other types of aerosols and a concrete inference needs further investigation. The highest fractional contribution of BC in winter and lowest in the monsoon season was reported from Anantapur [37] which is quite opposite to the seasonal variation of BC to $PM_{2.5}$ ratio of the present study.

4.3 Columnar Aerosols Versus Surface Aerosols

Correlation and regression analyses between columnar AOD and surface level concentration of $PM_{2.5}$ are conducted for the pre-monsoon, monsoon, post-monsoon, and winter seasons, with the findings being displayed in Fig. 7a–d, respectively. Weak correlation was observed between AOD and $PM_{2.5}$ in all four seasons with very low values of correlation coefficients. The observed poor correlation between AOD and $PM_{2.5}$ indicates the presence of aerosols at higher altitudes over the study region. Coefficients of regression analysis show the highest value of intercept (0.69) in pre-monsoon, which indicates the presence of more elevated aerosols during this season. The correlation coefficients of the relationship between ground-level $PM_{2.5}$

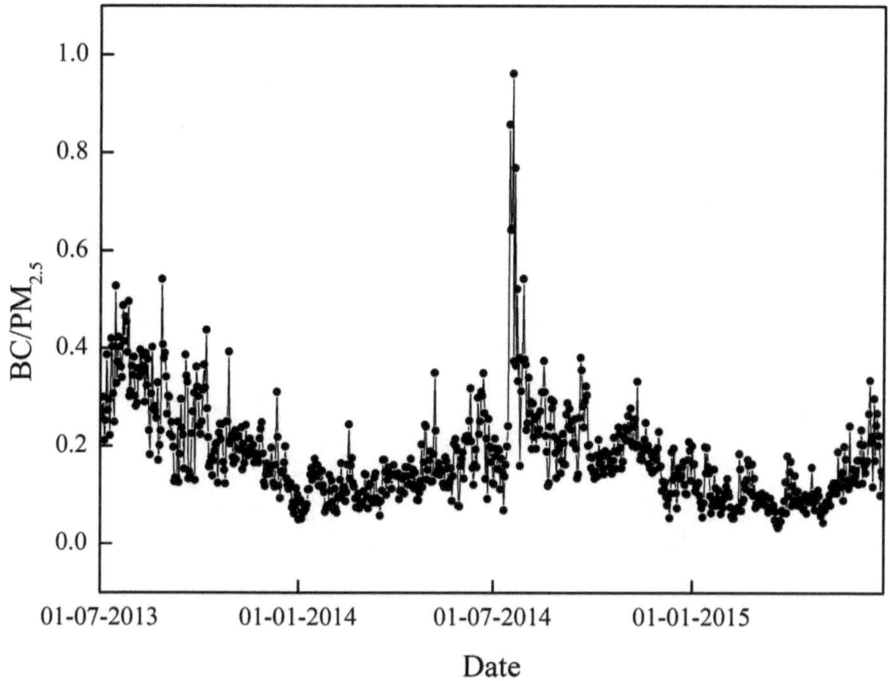

Fig. 6 Time series of BC to $PM_{2.5}$ mass ratio over Tezpur during July 2013–June 2015 (*Source* Author)

and MODIS AOD 550 were 0.69, 0.60, and 0.73 for spring, summer, and autumn, respectively, in Beijing during 2012–13 [38]. Correlation analysis between AOD and $PM_{2.5}$ yielded good results during summer and fall in different places of the US while showing poor correlation during winter [39].

4.4 Elevated Aerosols During Pre-monsoon

Further confirmation of the presence of aerosols at higher altitudes over the study region is provided by the vertical distribution of aerosols from CALIOP observations made in 2014 and 2015 during the pre-monsoon seasons. In Fig. 8, daytime (star) and nighttime (circle) vertical profiles of aerosol extinction coefficients at 532 nm during the pre-monsoon are displayed. Solid lines indicate the mean values during the season and horizontal bars represent corresponding standard errors. Presence of aerosols can be observed at higher altitudes up to ~4 km in day and night during the season. Extinction coefficients are observed as high as 0.2 km^{-1} in the daytime, whereas it is up to 0.3 km^{-1} at night at altitudes slightly below 1 km. Elevated aerosols are so important as the impacts of aerosol are more intense at higher altitudes, mostly

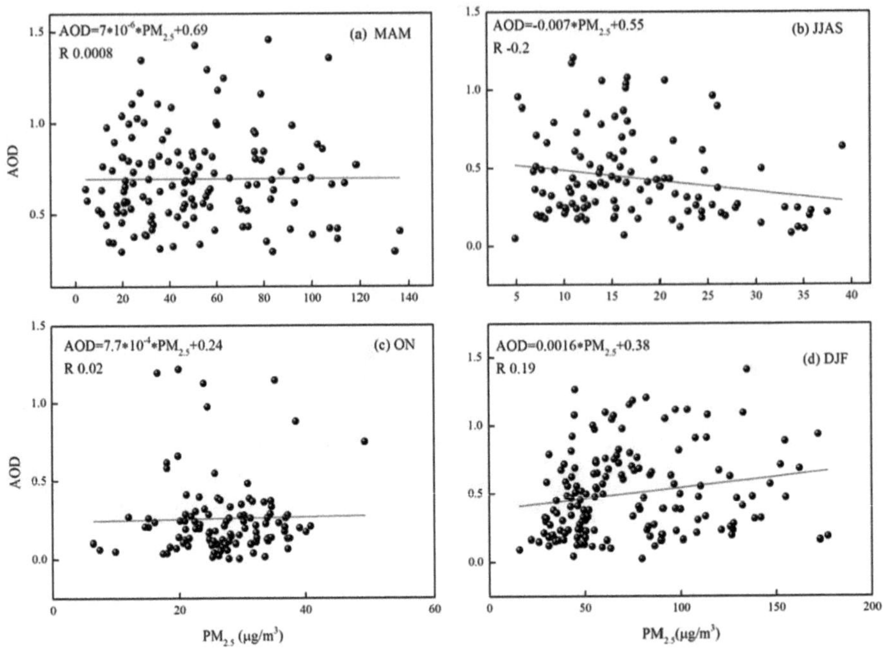

Fig. 7 Correlation between $PM_{2.5}$ and AOD during: **a** pre-monsoon, **b** monsoon, **c** post-monsoon, and **d** winter. The grey solid line represents the least square linear fit (*Source* Author)

as a result of the quick drop in air density. Presence of aerosols, particularly BC, at higher altitudes can increase the rate of global warming.

5 Conclusions and Way Forward

The present study examines columnar and surface concentrations of aerosols in different seasons over Northeast India using satellite observations and ground-based measurements. The pre-monsoon season recorded the highest values of columnar AOD while the post-monsoon recorded the lowest values, whereas surface concentrations of $PM_{2.5}$, PM_{10}, and BC are found higher in winter. The fractional contribution of BC in the total mass concentration of aerosols at the surface level is highest in monsoon. Correlation and regression analysis between columnar AOD and surface mass concentration of aerosols reveal a significant amount of aerosols at higher altitudes over Northeast Indian regions, especially during pre-monsoon. The presence of elevated aerosols during pre-monsoon is further confirmed using vertical profiles of aerosol extinction coefficient from satellite observations. During pre-monsoon, aerosol concentration is found significant at altitudes up to 4 km over the study region. Elevated aerosols could significantly raise the rate of global warming.

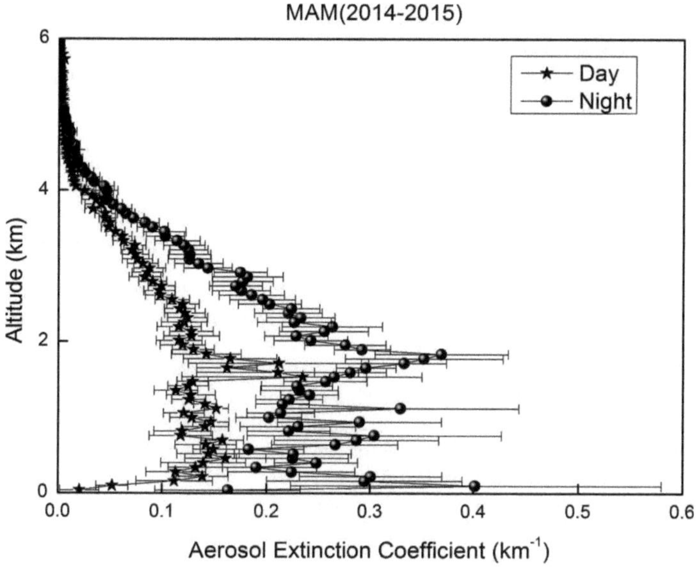

Fig. 8 Aerosol extinction coefficients at 532 nm from CALIPSO over the study region during pre-monsoon season of 2014–2015 in day (star) and night (circle). Solid lines indicate the mean values during the season and horizontal bars represent corresponding standard errors (*Source* Author)

Higher surface concentrations of PM and BC during the critical winter season can pose serious health risks. The findings of this study can serve as a baseline information to make policy decisions related to air quality management, public health strategies, or climate change mitigation in Northeast India.

Continuous monitoring of aerosols along with meteorological parameters should be carried out for years in future to develop mathematical models and to study the aerosol trend in this region. Aerosol concentrations can feed as one of the several inputs required to run forecasting models like WRF-Chem (Weather Research and Forecasting (WRF) model coupled with Chemistry). Spectral or Fourier analysis of long-term aerosol data series should be performed to identify the typical periodicities caused by anthropogenic and meteorological influences. BC study for long term can help in understanding its impact on regional climate change, thereby, finding solutions to reduce air pollution and build a safe, sustainable environment.

References

1. Pöschl U (2005) Atmospheric aerosols: composition, transformation, climate and health effects. Atmos Chem 44:7520–7540
2. Lagzi I, Mészáros R, Gelybó G, Leelőssy Á (2013) Aerosol particles. In: Atmospheric chemistry. Eötvös Loránd University, Hungary

3. IPCC (Intergovernmental Panel on Climate Change) (2014) Climate change 2013: the physical science basis. Cambridge University Press, New York
4. Boucher O (2015) Atmospheric aerosols. atmospheric aerosols: properties and climate impacts. Springer, Netherlands. https://doi.org/10.1007/978-94-017-9649-1_2
5. Tomasi C, Lupi A (2017) Primary and secondary sources of atmospheric aerosol. In: Atmospheric aerosols: life cycles and effects on air quality and climate. Wiley-VCH Verlag GmbH & Co., KGaA
6. Prospero JM, Charlson RJ, Mohnen V, Jaenicke R, Delany AC, Moyers J, Zoller W, Rahn K (1983) The atmospheric aerosol system: an overview. Rev Geophys Space Phys 21(7):1607–1629
7. Whitby K (1978) The physical characteristics of sulfur aerosols. Atmos Environ 12:135–159
8. Geerts B. Aerosol and climate. http://wwwdas.uwyo.edu/~geerts/cwx/notes/chap02/aerosol&climate.html. Accessed 26 Mar 2018
9. Myhre G, Myhre CEL, Samset BH, Storelvmo T (2013) Aerosols and their relation to global climate and climate sensitivity. Nat Educ Knowl 4(5):7
10. Verheggen B, Weijers EP (2010) Climate change and the impact of aerosol. ECN-E-09-095
11. Zhang R, Ho K, Shen Z (2012) The role of aerosol in climate change, the environment, and human health. Atmos Oceanic Sci Lett 5(2):156–161
12. Fierro M. Particulate matter. file:///Ll/Air_updates/particulate matter-singspace.html. Accessed 16 Aug 2012
13. Particulate matter emissions. USEPA Report on the Environment. http://www.epa.gov/roe/. Accessed 16 May 2017
14. Balakrishnaiah G, Kumar KR, Reddy BSK, Gopal KR, Reddy RR, Reddy LSS, Narasimhulu K, Ahammed YN, Balanarayana C, Moorthy KK, Babu SS (2011) Characterization of PM, PM10 and PM2.5 mass concentrations at a tropical semi-arid station in Anantapur, India. Indian J Radio Space Phys 40:95–104
15. Tiwari S, Srivastava AK, Bisht SD, Bano T, Singh S, Behura S, Srivastava MK, Chate DM, Padmanabhamurty B (2009) Black carbon and chemical characteristics of PM10 and PM2.5 at an urban site of North India. J Atmos Chem 62:193–209
16. Black carbon: a short lived climate forcer; cannot substitute CO_2 mitigation strategies. Centre for Science and Environment. https://www.cseindia.org/. Accessed 18 Sept 2013
17. Ramanathan V, Carmichael G (2008) Global and regional climate changes due to black carbon. Nat Geosci 1:221–227
18. Black carbon research and future strategies: reducing emissions, improving human health and taking action on climate change. In USEPA factsheet, www.epa.gov/research. Accessed 1 May 2013
19. Menon S, Hansen J, Nazarenko L, Luo Y (2002) Climate effects of black carbon aerosols in China and India. Science 297:2250–2253
20. Jacobson MZ (2010) Short-term effects of controlling fossil-fuel soot, biofuel soot and gases, and methane on climate, arctic ice, and air pollution health. J Geophys Res 115
21. Koch D, Del Genio AD (2010) Black carbon semi-direct effects on cloud cover: review and synthesis. Atmos Chem Phys 10:7685–7696
22. Chelani AB (2013) Statistical characteristics of ambient PM2.5 concentration at traffic site in Delhi: source identification using persistence analysis and nonparametric wind regression. Aerosol Air Qual Res 13(6):1768–1778
23. http://online.assam.gov.in/web/guest/assamgazetter?webContentId=172953. Accessed 30 Sept 2023
24. Kaufman YJ, Tanré D, Remer LA, Vermote EF, Chu A, Holben BN (1997) Operational remote sensing of tropospheric aerosol over land from EOS moderate resolution imaging spectroradiometer. J Geophys Res 102:17051–17068
25. Remer LA, Kleidman RG, Levy RC, Kaufman YJ, Tanré D, Mattoo S, Martins JV, Ichoku C, Koren I, Yu H, Holben BN (2008) Global aerosol climatology from the MODIS satellite sensors. J Geophys Res 113(D14)

26. Murari V, Kumar M, Barman SC, Banerjee T (2015) Temporal variability of MODIS aerosol optical depth and chemical characterization of airborne particulates in Varanasi, India. Environ Sci Pollut Res 22:1329–1343
27. Winker DM, Hunt WH, McGill MJ (2007) Initial performance assessment of CALIOP. Geophys Res Lett 34
28. Chung A, Chang DPY, Kleeman MJ, Perry KD, Cahill TA, Dutcher D, McDougall EM, Stroud K (2001) Comparison of real-time instruments used to monitor airborne particulate matter. J Air Waste Manag Assoc 51(1):109–120
29. Rosen H, Hansen AD, Dod RL, Novakov T (1980) Soot in urban atmospheres: determination by an optical absorption technique. Science 208(4445):741–744
30. Singh RP, Dey S, Tripathi SN, Tare V, Holden B (2004) Variability of aerosol parameters over Kanpur, Northern India. J Geophys Res 109
31. Kumar TK, Gadhavi H, Jayaraman A, Sai Suman MN, Rao SVB (2013) Temporal and spatial variability of aerosol optical depth over South India as inferred from MODIS. J Atmos Solar Terr Phys 94:71–80
32. Kant Y, Patel P, Mishra AK, Dumka UC, Dadhwal VK (2012) Diurnal and seasonal aerosol optical depth and black carbon in the Shiwalik hills of the North Western Himalayas: a case study of the Doon Valley, India. Int J Geol Earth Environ Sci 2(2):173–192
33. Bhuyan PK, Gogoi MM, Moorthy KK (2005) Spectral and temporal characteristics of aerosol optical depth over a wet tropical location in northeast India. Adv Space Res 35:1423–1429
34. Husain L, Dutkiewicz VA, Khan AJ, Ghauri BM (2007) Characterization of carbonaceous aerosols in urban air. Atmos Environ 41:6872–6883
35. Mahapatra PS, Panda S, Das N, Rath S, Das T (2014) Variation in black carbon mass concentration over an urban site in the eastern coastal plains of the Indian sub-continent. Theoret Appl Climatol 117:133–147
36. Ramachandran S, Rajesh TA (2007) Black carbon aerosol mass concentrations over Ahmedabad, an urban location in western India: comparison with urban sites in Asia, Europe, Canada, and the United States. J Geophys Res 112
37. Reddy BSK, Kumar KR, Balakrishnaiah G, Rama Gopal K, Reddy RR, Reddy LSS, Ahammed YN, Narasimhulu K, Moorthy KK, Babu SS (2012) Potential source regions contributing to seasonal variations of black carbon aerosols over Anantapur in Southeast India. Aerosol Air Qual Res 12:344–358
38. Guo Y, Feng N, Christopher SA, Kang P, Zhan FB, Hong S (2014) Satellite remote sensing of fine particulate matter (PM2.5) air quality over Beijing using MODIS. Int J Remote Sens 35(17):6522–6544
39. Zhang H, Hoff RM, Engel-Cox JA (2009) The relation between moderate resolution imaging spectroradiometer (MODIS) aerosol optical depth and PM2.5 over the United States: a geographical comparison by U.S. environmental protection agency regions. J Air Waste Manag Assoc 59(11):1358–1369

Spatio-Temporal Comparative Study of Aerosol Optical Depth, Aerosol Index, and Planetary Boundary Layer Height During and After COVID-19 Lockdown Over Northeastern Region of India

Divyanshu Saini and Dilip H. Lataye

1 Introduction

The COVID-19 pandemic, caused by the novel coronavirus SARS-CoV-2, has had a profound and far-reaching impact on global health. It was officially declared a pandemic by the World Health Organization (WHO) on 11th March 2020. The rapid urbanization and Industrialization of cities around the world have led to a major concern i.e. air pollution. So, this study outlines strategies to improve air quality in Indian cities, including public transport, emissions regulations, and waste management, emphasizing the importance of public awareness and integration with broader development policies for effective air quality control. To control the air pollution, researchers, analyze global air quality trends from 2010 to 2016 in the context of efforts to reduce air pollution, using a comprehensive model called DIMAQ to assess health impacts. This study found that during this period, 55.3% of the world's population experienced increased $PM_{2.5}$ exposures, with regional variations in the trend [1].

A study conducted for 13 years using satellite data on Aerosol Optical Depth in India identified that the Indo-Gangetic Plains had the highest annual aerosol levels, with variations influenced by factors like humidity and rainfall [2]. Another research also observed rising trends in aerosol levels during winter and post-monsoon seasons, potentially linked to increased human activity. Some analyzed tropospheric Aerosol Optical Depth (AOD) in India, finding stable AOD levels year-round, with peak values in summer. Their study revealed higher AOD levels in the SMOG inventory compared to the GC inventory, primarily due to black and organic carbon, and showed

D. Saini (✉) · D. H. Lataye
Visvesvaraya National Institute of Technology, Nagpur, Maharashtra, India
e-mail: saini.divyanshu12@gmail.com

V. Chettry et al. (eds.), *Geoinformatics for Hilly Regions*, Advances in 21st Century Human Settlements, https://doi.org/10.1007/978-981-95-0156-4_9

spatial AOD variations across India with the Indo-Gangetic Plain having the highest AOD [3].

The temporal and spatial distribution of four aerosol types in East Asia using MODIS-OMI and Four Channel Algorithms revealed consistent aerosol type distribution in the region, with agreement ranging from 32 to 81% [4]. Further, impact of global aerosol fluctuations on climate in India's Western Coastal region (2014–2017) was also attempted. They found that increased aerosol levels led to more rainfall due to changes in cloud properties and the presence of dust from Arabian Desert regions. Regional climate-chemistry model was used to study the relationship between anthropogenic aerosols and the East Asian Summer Monsoon (EASM) [5]. It was found that aerosols, primarily sulfates and black carbon, have a significant cooling effect on the EASM, but the warming impact of black carbon partially offsets this effect, and the influence is stronger in years with weaker winds at 850 hPa, affecting aerosol dispersion in the lower troposphere.

Long-term aerosol optical properties in the Tibetan Plateau were studied using a sun photometer and satellite data, revealing increasing Aerosol Optical Depth in recent decades [6]. MISR AOD_555 nm data for Western India from 2000 to 2016 was used to identify that El Niño years are linked to higher aerosol levels over cities like Surat, Mumbai, and Jodhpur, compared to La Niña years [7]. In 2019, PBLH over Gwal Pahari, New Delhi was studied for nearly a year using a multiwavelength Ra-man lidar and a modified WCT approach [8]. The method showed good efficiency ($r = 0.9$) in attributing PBLH despite aerosol variations, with satisfying results when compared to CALIPSO observations ($r = 0.8$), and they observed a slight weakening of the seasonal PBLH cycle compared to long-term climate records due to changing weather conditions.

The impact of aerosol vertical distribution on thermodynamic stability and boundary layer development in Beijing. They found that different aerosol vertical structures affect the interaction with the planetary boundary layer, diurnal PBL height variations, and PM2.5 levels, highlighting the significance of considering aerosol vertical distribution in modeling this interaction [9]. The study by Saha examines COVID-19's impact on the atmospheric boundary layer (ABL) over an urban area in western India. It is found that during the national lockdown in 2020, there was an improvement in air quality and a decrease in ABL height, which was attributed to rainy incidents and supported by ERA5 reanalysis data [10]. Further, regional meteorology, the topography of the area, and anthropogenic pollutants are all linked to air pollution levels [11].

The World Air Quality Report (2019) [12–14] lists Northeastern Indian states as key states that come from not-so-concerned parts of India as far as air pollution. Therefore, in this research, the analysis of the air pollutants and aerosols has been performed for the Northeastern State of India, i.e., Assam, Arunachal Pradesh, Manipur, Mizoram, Meghalaya, Tripura, and Nagaland.

2 Methodology

2.1 Study Area

In this research, the analysis of the PBLH and AOD has been carried out for the Northeastern State of India that named Assam (26.2006 °N, 92.9376 °E), Arunachal Pradesh (28.2180 °N, 94.7278 °E), Meghalaya (25.4670 °N, 91.3662 °E), Mizoram (23.1645 °N, 92.9376 °E), Manipur (24.6637 °N, 93.9863 °E), Tripura (23.5639 °N, 91.6761 °E), and Nagaland (26.1584° N, 94.5624° E). The location map of all the Northeastern states of India considered for the analysis is shown in Fig. 1.

Fig. 1 Study area map of Northeastern India with seven states (*Source* Authors)

Table 1 Properties of satellite data products of aerosols and PBLH obtained from Giovanni

Satellite/sensor	Parameter	Spatial resolution	Wavelength	Product level
MODIS-Terra	AOD	10	550 nm	MOD08_D3_v6.1
OMI-Aura	AI	10	UV	OMTO3_v003
MERRA-2 model	PBLH	0.50 × 0.625	Microwave and hyperspectral infrared	M2TMNXFLX_ 5.12.4

2.2 *Data Collection*

For this analysis, three Earth Observation System (EOS) satellites, namely MODIS Terra Level 3 (MOD08_D3 v6.1), OMI Aura (OMTO3d v003), and the MERRA-2 model (M2TMNXFLX v5.12.4), were employed to analyze Aerosol Optical Depth (AOD), Aerosol Index (AI), and planetary boundary layer height (PBLH), respectively. MODIS Terra provides daily AOD data at a wavelength of 550 nm, OMI Aura supplies daily AI data at UV wavelengths, and the MERRA-2 model offers monthly observations of PBLH. These datasets can be accessed through Giovanni, an online application available from the NASA Goddard Earth Science Data and Information Services Center (GES DISC), which provides a user-friendly platform for accessing and visualizing remote sensing data related to Earth science. These data were utilized to quantify changes in aerosol levels and the planetary boundary layer during and after the COVID pandemic lockdown period. Additional information on these three satellite data products is provided in Table 1.

2.3 *PBLH, AOD, and AI*

Planetary Boundary Layer Height (PBLH) is a meteorological term that refers to the height or thickness of the planetary boundary layer, which is the lowest part of the Earth's atmosphere. The planetary boundary layer is characterized by turbulent mixing of air and is directly influenced by the Earth's surface, such as terrain, vegetation, and human activities.

Aerosol Optical Depth (AOD) is a measure of the extent to which aerosols, such as dust, smoke, and fine particulate matter, in the Earth's atmosphere can absorb and scatter incoming solar radiation. It quantifies the attenuation of light caused by aerosol particles in a column of air. AOD is typically measured along the vertical path from the top of the Earth's atmosphere to the surface.

The Aerosol Index (AI) is a quantitative measure used in atmospheric science and remote sensing to assess and quantify the presence and concentration of aerosol particles in the Earth's atmosphere. Aerosols are tiny solid or liquid particles suspended in the air, which can include dust, smoke, pollution, sea salt, and various other particulate matter. The AI is typically derived from satellite or ground-based observations

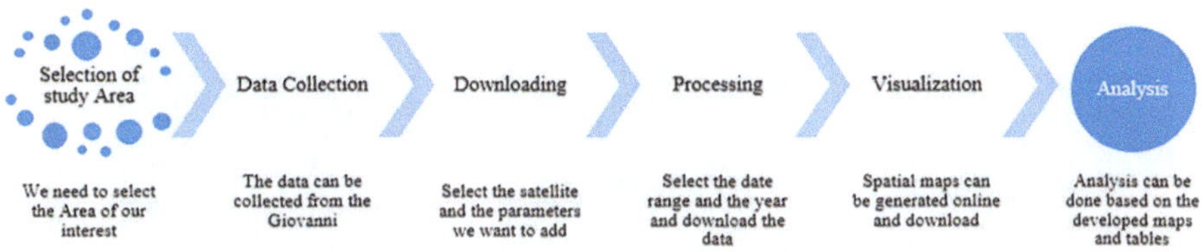

Fig. 2 Methodological flowchart (*Source* Authors)

and is used to provide information about the atmospheric aerosol load and its optical properties. The methodological flowchart is being added in Fig. 2.

3 Result and Discussion

This section provides a comprehensive summary of the study's findings, delving deep into the generated values. The results are presented through spatio-temporal maps, tables displaying monthly average values, and a correlation map between AOD and PBLH.

3.1 Spatial and Temporal Variation of AOD

Spatial variation of AOD at 550 nm for Northeastern India for the months of October, November, and December during the first COVID-19 lockdown, and the post-COVID period (2021) are shown in Fig. 3. It is observed that there is an increasing trend in AOD from 2019 to 2021. There was an extremely strict lockdown in April 2019 which was the initial phase of COVID-19 lockdown when all activities were prohibited except medical emergencies which resulted in a decrease in the AOD values in the atmosphere. However, in the state-wide second-wave lockdown, some activities were allowed presuming that the individuals were more aware of COVID-19, which led to steadily rising AOD values. This clearly illustrates how anthropogenic activities are mostly responsible for the degradation of air quality.

As shown in Fig. 2, Tripura shows the highest AOD values of 0.35(2019), 0.42(2020), and 0.41(2021) among all Northeastern states, while Arunachal Pradesh shows the lowest AOD values of 0.17(2019), 0.15(2020), 0.15(2021) for the month of December over the period of three years.

A similar spatial variation is shown in Fig. 3 for the November months of 2019, 2020, and 2021. AOD values in all parts increased during the second wave of the lockdown. In 2021, AOD values in Assam and Tripura are high. This shows that industrial activities and vehicular emissions are increasing rapidly. In November 2019, Arunachal Pradesh had the lowest AOD value. It shows that an increase in

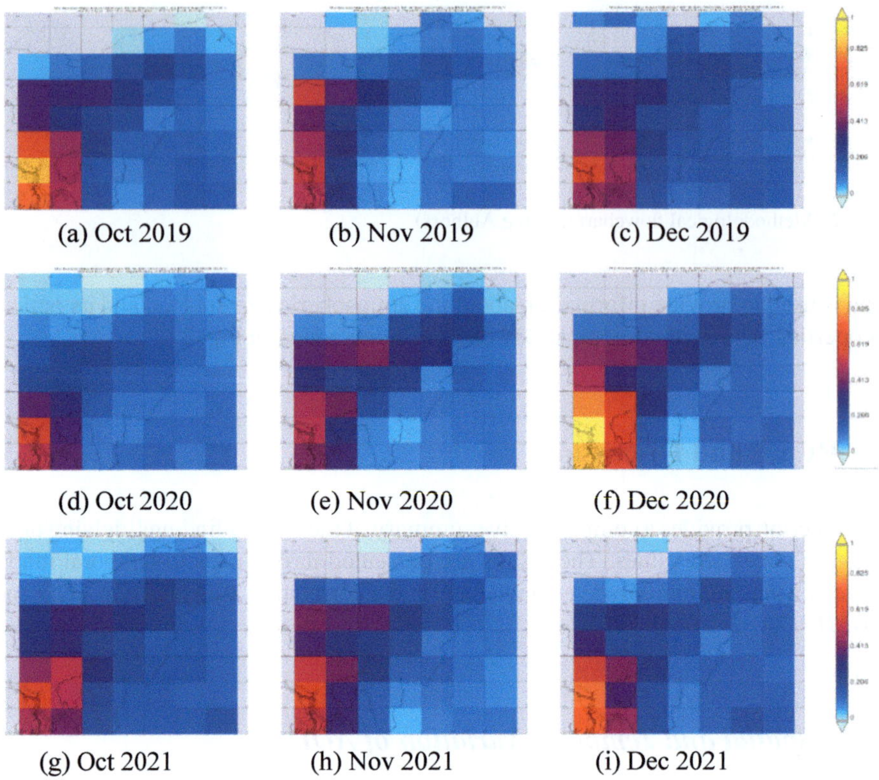

Fig. 3 Distribution and fluctuations in the monthly mean AOD at 550nm for the months of October, November, and December in the years 2019 [**a**, **b**, and **c**], 2020 [**d**, **e**, and **f**], and 2021 [**g**, **h**, and **i**], respectively (*Source* Authors)

anthropogenic and industrial activities causes a rise in air pollution which needs to be monitored and controlled.

Figure 3 shows the statistical representation of the monthly mean AOD variation at 550 nm for all selected states in India. The majority of the states show a clear increase in values of AOD for 2021. It provides a clear picture of how aerosol loading varies in relation to emissions caused by anthropogenic activities, industrial activities, and vehicular emissions.

3.2 Spatial and Temporal Variation of PBLH

Month-wise change in PBLH for the 7 states is shown in Fig. 4. The monthly data have been examined in order to understand the spatial and temporal variation of PBLH. As shown in Fig. 3a, b, c, d, e, f, g, h, and i among all the 3 years for October,

November, and December month, Nagaland shows the highest PBLH (1366.85m during 2020, 1138.85 m during 2019, and 1065.43 m during 2021) while Tripura shows the lowest PBLH (474.27 m during 2021 and 451.25 m during 2020) value. Nearly all states have seen a small rise from 2020 to 2021.

PBLH for the month of December is shown in Fig. 4 (c), (f), and (i). The results show that except for Tripura and Mizoram, PBLH values are falling from 2020 to 2021 for almost all states. However, some states have a declining value from 2020 to 2021, while the PBLH values have increased in other states (Table 2). This could occur as the absorbing aerosol warms the lower atmosphere, lowering the inversion and causing the PBLH to rise.

Results shown in Fig. 4 (a), (b), and (c) depict the spatial variation for October which reveals that all cities have a minor percentage of variation in 2021 as given in Table 3, with Tripura and Mizoram implying a decreasing trend while the remaining states have an increasing trend.

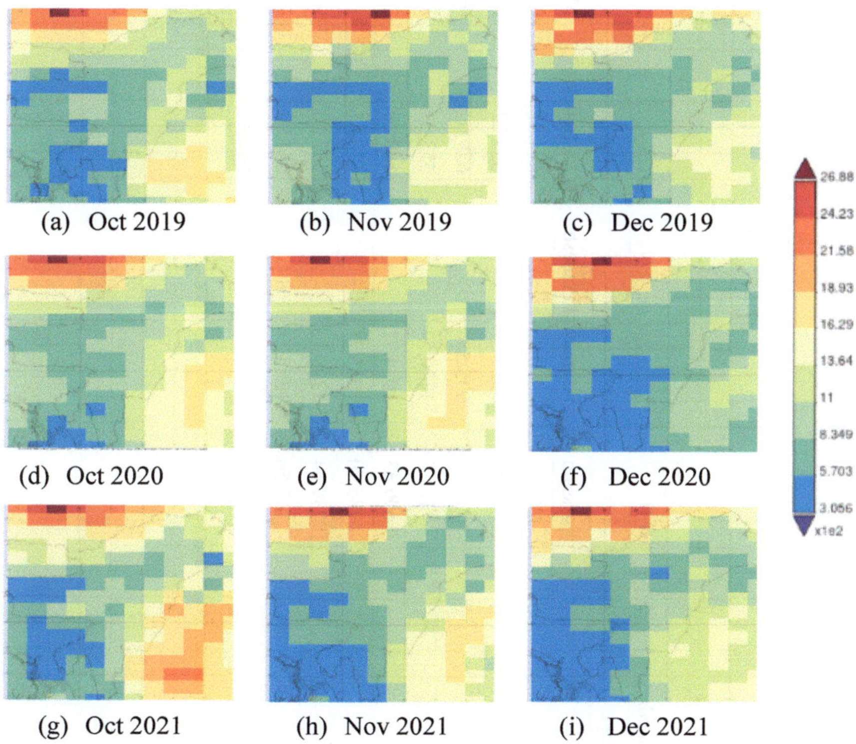

Fig. 4 Distribution and fluctuations in the monthly mean PBLH for the months of October, November, and December in the years 2019 [**a**, **b**, and **c**], 2020 [**d**, **e**, and **f**], and 2021 [**g**, **h**, and **i**], respectively (*Source* Authors)

Table 2 Planetary boundary layer height PBLH

Planetary boundary layer height PBLH							
Months	Assam	Arunachal Pradesh	Meghalaya	Manipur	Mizoram	Tripura	Nagaland
Oct-19	743.68	1000.86	654.21	906.61	605.33	582.89	883.76
Oct-20	679.45	892.92	639.22	855.54	619.68	531.47	785.41
Oct-21	727.96	935.58	629.02	964.32	644.65	553.71	807.21
Nov-19	716.18	1120.54	553.27	840.46	512.90	613.97	933.00
Nov-20	699.83	1034.71	560.29	764.55	672.54	634.66	946.36
Nov-21	785.70	1085.71	610.67	1064.31	544.14	474.27	997.44
Dec-19	739.44	1200.16	572.50	914.39	577.09	579.08	1138.18
Dec-20	762.56	1285.71	561.99	792.79	429.27	451.32	1366.85
Dec-21	729.05	1049.92	607.53	936.79	656.05	544.82	1065.43

Table 3 Aerosol index

Aerosol index							
Months	Assam	Arunachal Pradesh	Meghalaya	Manipur	Mizoram	Tripura	Nagaland
Oct-19	−7E + 29	−6E + 29	−7E + 29	−9E + 29	−1E + 30	−9E + 29	−1E + 30
Oct-20	−5E + 29	−4E + 29	−2E + 29	−7E + 29	−7E + 29	−4E + 29	−1E + 30
Oct-21	−5E + 29	−9E + 29	−5E + 29	−7E + 29	−8E + 29	−6E + 29	−1E + 30
Nov-19	−1E + 29	−2E + 29	−3E + 29	−6E + 29	−8E + 29	−7E + 29	−1E + 30
Nov-20	−4E + 28	−4E + 29	−4E + 29	−7E + 29	−8E + 29	−7E + 29	−1E + 30
Nov-21	−3E + 29	−8E + 29	−4E + 29	−1E + 30	−7E + 29	−6E + 29	−1E + 30
Dec-19	−1E + 29	−1E + 29	−3E + 29	−4E + 29	−6E + 29	−5E + 29	−1E + 30
Dec-20	−2E + 29	−7E + 29	−6E + 29	−6E + 29	−6E + 29	−7E + 29	−1E + 30
Dec-21	−3E + 29	−1E + 30	−1E + 30	−1E + 30	−1E + 30	−1E + 30	−1E + 30

3.3 Aerosol Index (AI)

AI believed to be indicative of UV-absorbing aerosols such as soot and desert sand, displayed intriguing dynamics over the course of various events. Notably, during the initial COVID-19 lockdown, many states reported consistently low daily AI values. However, in 2020, AI values began to surge, reaching a peak, and by 2021, they surpassed. This significant increase can be attributed to a series of factors: The initiation of numerous activities, and a second wave of lockdown followed by a resurgence in industrial operations and vehicular traffic. These activities contributed to higher emissions and, consequently, an augmented presence of carbonaceous matter in the atmosphere.

Notably, specific northern states of the Northeastern region such as Arunachal Pradesh, Assam, and Nagaland exhibited notably higher AI values. This phenomenon can be explained by various factors, including local wind directions, wind speed, and humidity levels in these regions. Additionally, during the second COVID wave, when certain activities resumed, AI values showed an upward trend in October and December 2021, with a slight decline in November 2021.

3.4 Spatial Correlation Between PBLH and AOD

Figure 5 that illustrates the relationship between AOD and PBLH through spatial correlation tool for the period from October to December 2022. The map reveals that the southern region of Northeastern India and certain areas of Arunachal Pradesh display a positive correlation, while the majority of locations exhibit a negative correlation. Notably, several states such as Tripura, Mizoram, Manipur, Meghalaya, and some parts of Nagaland exhibit a significant negative connection. This negative correlation suggests that as AOD increases, the PBLH tends to decrease. This relationship can be explained by the fact that higher aerosol concentrations enhance albedo and slow down the development of the Planetary Boundary Layer (PBL). On the other hand, some states such as Assam, some parts of Manipur, and Nagaland display a positive correlation. This positive correlation may be attributed to the presence of absorptive aerosols, which promote PBL formation. These findings suggest that there is likely a higher level of anthropogenic activity in these areas. Additionally, integrating socio-economic and land-use data could help identify the specific anthropogenic sources contributing to aerosol pollution and their impacts on atmospheric conditions. By clarifying the observational nature of correlations and emphasizing the need for additional research to establish causation, this study contributes to advancing our understanding of the complex interplay between aerosols and atmospheric dynamics.

By adopting a more regionally focused approach and accounting for local environmental and socio-economic factors, policymakers and stakeholders can better tailor air quality management strategies to address the specific challenges and priorities of each region, ultimately contributing to global efforts toward achieving cleaner and healthier atmospheres. The research findings inform clean air action plans by targeting key emission sources, guiding urban planning to reduce exposure,

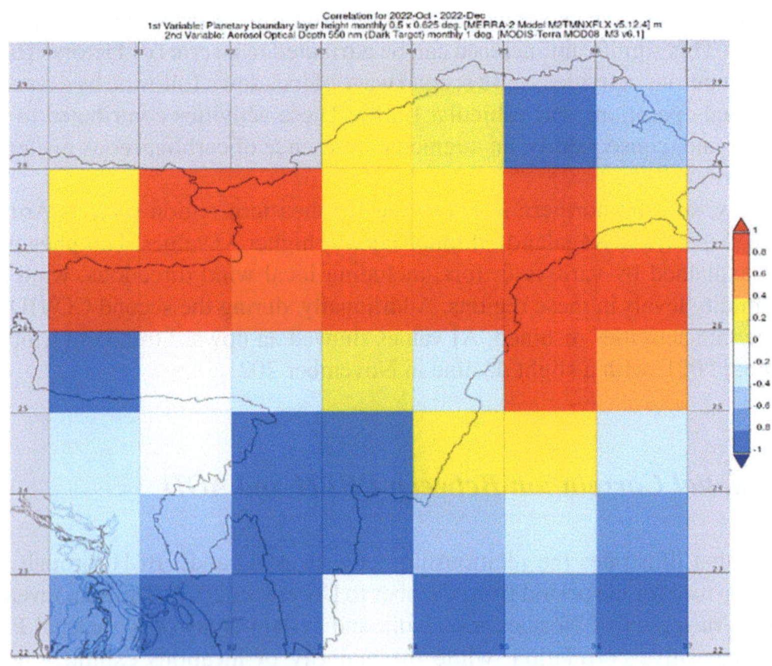

Fig. 5 Spatial correlation map of AOD and PBLH where legend showing correlation coefficient (*Source* Authors)

supporting public health interventions, strengthening air quality regulations, and prioritizing resource allocation for impactful initiatives. This ensures policies are data-driven and effective in improving air quality and health.

4 Conclusion and Recommendations

The analysis conducted to assess spatio-temporal variations in Aerosol Optical Depth (AOD), Planetary Boundary Layer Height (PBLH), and their relationship with Aerosol Index (AI) during and after the COVID-19 lockdown has yielded the following key findings. The results indicate that the concentration of aerosols in the atmosphere, as measured by AOD and AI data from MODIS Terra and OMI Aura, was higher during the second wave of the COVID-19 lockdown in 2021 compared to the initial lockdown in 2020. Moreover, aerosol concentration continued to increase during the post-lockdown period in 2021. This increase in aerosol concentration suggests that the relaxation of lockdown restrictions led to a gradual resurgence in anthropogenic and industrial activities, resulting in higher pollutant emissions and subsequently greater aerosol loading. This research offers valuable insights into the significant role of human-generated aerosols in deteriorating air quality. In general, AI values are positive for absorbing aerosols and negative for non-absorbing aerosols. Over the years, the study identified 8 states that consistently displayed positive AI values. The current research demonstrates that these 8 states are predominantly influenced by absorbing aerosols such as dust, smoke, and emissions from biomass burning, indicating higher levels of anthropogenic emissions. The Planetary Boundary Layer Height (PBLH) exhibited higher values during the second COVID-19 wave lockdown and the post-lockdown phase. PBLH values demonstrated a steady increase from October 2019 to December 2021, following the relaxation of activities after the initial lockdown in March 2020. Notably, PBLH values were at their lowest in October 2021 following the second wave of COVID-19, coinciding with a complete lockdown in many cities. It can be inferred that the rise in PBLH values is closely linked to the presence of powerful absorptive aerosols, such as smoke, dust, and soot, in the lower atmosphere, driven by AI values. Instead of suppressing PBLH, these absorbing aerosols at lower altitudes warm the atmosphere, leading to an elevation in PBLH.

Traditionally, AOD and PBLH demonstrate an inverse relationship in monthly data. However, in the current study, both AOD and PBLH values exhibited an upward trend over the three years examined. The spatial correlation map indicates that AOD consistently exhibits a negative correlation with PBLH. However, the type of aerosols present in the area can influence the development of PBLH. The COVID-19 lockdown had a substantial impact on human activities, resulting in reduced levels of atmospheric pollutants. As a consequence, many cities in India experienced an improvement in their Aerosol Index. However, the resumption of industrial operations following the second lockdown led to a subsequent increase in pollutant concentrations, resulting in a decline in air quality. This ongoing trend has the potential to impede the nation's ability to achieve its air quality targets in the future. Furthermore, aerosol particles also affect another factor known as PBLH. This study will provide insights for the development of action plan for the clean air program and will assist decision-makers and policy developers in determining the most effective strategies to reduce the anthropogenic contributions impacting aerosol concentrations in the atmosphere.

The study recommends the implementation of strict emission controls by enforcing industry standards and promoting cleaner technologies. It suggests enhancing monitoring systems through improved technologies and data-sharing mechanisms. Furthermore, sustainable practices such as renewable energy and sustainable agriculture should be encouraged, while air quality considerations must be integrated into urban planning through zoning and infrastructure development. Public awareness campaigns are essential to educate citizens about air quality and sustainable lifestyles. The study also advocates for fostering international collaboration to set common standards and address regional pollution, alongside investing in research to identify pollution sources and develop effective mitigation strategies. Implementing these recommendations can help reduce anthropogenic aerosol emissions and protect air quality for future generations.

The future scope of the study includes analysis of the interactions between different pollutants and their combined impact on human health and the environment. Additionally, investigating the effectiveness of specific air quality management strategies can help assess the impact of proposed policy measures. This could include evaluating the outcomes of emission reduction initiatives, urban planning adjustments, and public health interventions to determine their long-term effectiveness in improving air quality and health outcomes.

References

1. Guttikunda SK, Nishadh KA, Gota S, Singh P, Chanda A, Jawahar P, Asundi J (2019) Air quality, emissions, and source contributions analysis for the Greater Bengaluru region of India. Atmos Pollut Res 10(3):941–953
2. Mehta M (2015) A study of aerosol optical depth variations over the Indian region using thirteen years (2001–2013) of MODIS and MISR Level 3 data. Atmos Environ 109:161–170
3. David LM, Ravishankara AR, Kodros JK, Venkataraman C, Sadavarte P, Pierce JR, Millet DB (2018) Aerosol optical depth over India. J Geophys Res: Atmos 123(7):3688–3703
4. Kim J, Lee J, Lee HC, Higurashi A, Takemura T, Song CH (2007) Consistency of the aerosol type classification from satellite remote sensing during the Atmospheric Brown Cloud–East Asia Regional Experiment campaign. J Geophys Res: Atmos, 112(D22)
5. Kumar KP, Manjunatha BR (2021) Impact of Aerosols on Precipitation over Western Ghats. Aerosol Sci Eng 5:393–403
6. Wang Q, Wang Z, Zhang H (2017) Impact of anthropogenic aerosols from global, East Asian, and non-East Asian sources on East Asian summer monsoon system. Atmos Res 183:224–236
7. Zhu J, Xia X, Che H, Wang J, Cong Z, Zhao T, Zhang Y (2019) Spatiotemporal variation of aerosol and potential long-range transport impact over the Tibetan Plateau. China. Atmos Chem Phys 19(23):14637–14656

8. Nakoudi K, Giannakaki E, Dandou A, Tombrou M, Komppula M (2019) Planetary boundary layer height by means of lidar and numerical simulations over New Delhi. India. Atmos Meas Tech 12(5):2595–2610
9. Su T, Li Z, Li C, Li J, Han W, Shen C, Guo J (2020) The significant impact of aerosol vertical structure on lower atmosphere stability and its critical role in aerosol–planetary boundary layer (PBL) interactions. Atmos Chem Phys 20(6):3713–3724
10. Saha S, Sharma S, Chhabra A, Kumar P, Kondapalli NK, Kamat D, Lal S (2023) Atmospheric boundary layer over Ahmedabad, Western Indian region: impact of COVID-19 Nationwide Lockdown. Pure Appl Geophys 180(3):1113–1119
11. Pérez IA, García MÁ, Sánchez ML, Pardo N, Fernández-Duque B (2020) Key points in air pollution meteorology. Int J Environ Res Public Health 17(22):8349
12. Air IQ (2019) World air quality report region & city Pm2. 5 ranking. Index Quality Air
13. Moreira GDA, Oliveira APD, Codato G, Sánchez MP, Tito JV, Silva LAHE, Landulfo E (2022) Assessing spatial variation of PBL height and aerosol layer aloft in Sao Paulo megacity using simultaneously two lidar during Winter 2019. Atmosphere 13(4):611
14. Samuel C, Kumar RP, Gautam S (2023) Review of retrieval of aerosol optical depth to estimate particle concentration and its challenges based on spatiotemporal relationships by various spectroradiometer models. Geol J 58(12):4403–4416

Geoinformatics for Climate Change Vulnerability Assessment in the Western Himalayan Cities

Ranjeet Verma, Harsimran Kaur, and Amrita Dwivedi

1 Introduction

The Himalaya is the newly formed fold mountain range in the world. This mountain range is the result of the collision between the Indian and the Eurasian Plates. The Himalayan Mountain range consists of the loftiest and rising peaks of the world and is known as "Himadri, or the abode of snow." Ten of the 14 highest mountain peaks globally are in the Himalayas. These peaks are the origin point of ten major rivers and the primary source of fresh water for nearly 1.3 billion people in Asia (almost 20% of the world population) [1, 2]. This mountain range is geologically very complex. These mountain ranges are formed primarily of sedimentary and metamorphic rocks, uplifted and folded by the collision of the Eurasian and Indian plates, and are also home to several active faults, which make the region prone to earthquakes [3]. This newly formed mountain range stretches over 2,400 kms from the Brahmaputra Gorge in the east to the Indus Gorge in the west, with the width varying from 240 – 350 km from west to east. The entire Himalayas are divided into the following three broad regions from west to east: (i) The western Himalayan region (WHR), (ii) the central Himalayan region, and (iii) the eastern Himalayan region. The WHR is the largest region of the Indian Himalayas, which comprises three states: Jammu and Kashmir (J & K), Himachal Pradesh, and Uttarakhand" [4]. Human settlements in the Himalayan region date back thousands of years. Starting from nomadic and unstable settlements, the region gradually became densely populated. Over time, the population in the region grew, and new settlements were established

R. Verma · A. Dwivedi (✉)
Department of Humanistic Studies, Indian Institute of Technology (Banaras Hindu University), Uttar Pradesh, Varanasi, India
e-mail: amrita.hss@itbhu.ac.in

H. Kaur
Department of Architecture, Planning and Design, Indian Institute of Technology (Banaras Hindu University), Uttar Pradesh, Varanasi, India

V. Chettry et al. (eds.), *Geoinformatics for Hilly Regions*, Advances in 21st Century Human Settlements, https://doi.org/10.1007/978-981-95-0156-4_10

in higher and higher altitudes. The Himalayas are one of the most inhabited mountain ranges in the world and are home to more than 210 million people [5]. According to NITI Aayog, 2022, in the Indian Region of Himalaya (IRH), more than 50 million people live. The rapid population growth in the Himalayan region is putting pressure on natural resources and making the region vulnerable to extreme weather events and climate change. The landscapes of this region are highly vulnerable to natural hazards, which raises concerns about climate change impacts both now and in the future [6]. The People and communities of this region are already experiencing the impacts of climate change-induced vulnerability.

Mountain regions are highly vulnerable to climate change and natural hazards due to their unique physical, topographical, and ecological characteristics worldwide. They are home to many ecosystems, including forests, grasslands, alpine meadows, and glaciers [7]. Many of these ecosystems are highly sensitive to changes in precipitation and temperature [8]. The mountainous regions are to be more severely affected by climate change than other regions on the Earth at the same latitudes. Climate change has a more significant impact on the Himalayan mountain ecosystem, which in turn poses new threats to livelihoods for not only those living in this region but also those living in the adjacent Indo-Gangetic plains whose survival and development are directly dependent upon the goods and services provided by the Himalayan ecosystems [9], 10, 11]. Accordingly, it is vital to observe trends in climate variability and extreme events in mountainous regions in order to detect global climate changes at an early stage [12]. Climate change is a natural phenomenon already affecting our environment, livelihoods, and ecosystems and thus poses a significant challenge to the global community. Due to accelerated population growth and climate change in the Himalayan cities, the mountain environment is threatened and vulnerable to various natural hazards. The aim of this study is to give an overview of the application of Geoinformatics in climate change vulnerability assessment for the western Himalayan mountain cities. Climate change vulnerability assessment (CCVA) is a process of identifying and understanding the risks posed by climate change to people, infrastructure, and ecosystems and making these resilient [13]. Although climate vulnerability assessments (CVA) play a vital role in addressing future climate risks, the assessment process may also contribute to improving the management of current climate risks. CVA for mountainous cities is essential for understanding the risk of climate change and developing strategies to make cities climate resilient by considering climate change exposure, sensitivity, and adaptive capacity of mountainous cities. Geoinformatics plays a significant role in CVA. It combines remote sensing, geographical information systems (GIS), and GPS and provides various applications in climate change analysis and adaptation by quantifying various aspects of the climate system. It provides tools and technologies for collecting, analyzing, and visualizing spatial data.

2 Climate Change in the Western Himalayas

The western Himalayan region (WHR), spanning over 331,392 square kilometers, constitutes approximately one-tenth of India's landmass, yet hosts only 2.44% of the country's population [14]. This region is rich in biodiversity, forest cover, agricultural landscapes, glaciers, wetlands, and urbanized territory. The varied topography and climate of the region support diverse ecosystems, including forests, grasslands, and alpine meadows. The biodiversity of the WHR is under threat from several factors, including climate change, habitat loss, overpopulation, and overexploitation of resources. The region has a fragile and susceptible ecosystem experiencing the impacts of climate change at an alarming rate. Population growth and climate are the most pressing challenges facing the WHR. These two challenges are interconnected, as population growth puts additional stress on resources and ecosystems already impacted by climate change.

According to Diksha & Amit (2017), the Himalayan capital cities are experiencing exponential growth (830.92%) with rapid population increase (333.45%) from 1970 to 2017. The western and central Himalayan cities are experiencing rapid urban growth [15], and the urban population in western Himalayan states is growing rapidly. The WHR is the most densely populated and rapidly urbanizing mountain system globally [5]. For example, Uttarakhand is part of the WHR and one of the most rapidly urbanizing Himalayan states in terms of urban population and number of towns. In Uttarakhand, in 1951, there were 34 towns in the state, which became 40 in 1971, and in 2011, it reached 115. If we look at the census data, during the decade between 1971 and 2011, in Uttarakhand, the urban population increased from 16.39% of the total population to 20.7% in 1981, 22.97% in 1991, 25.59% in 2001, and 30% in 2011 (refer to Table 10.1). Similar population growth is faced by the other WHR states such as Himachal Pradesh, Jammu and Kashmir, and Leh-Ladakh (earlier, it was part of Jammu and Kashmir state). The need for an urban land-use policy is the main reason for rapid and unplanned urbanization in the environmentally sensitive Himalayan region. This lack of policies governing urban land use has led to a shift in urbanization from geologically stable mid-slopes and ridges to environmentally more sensitive higher elevations. The growing urban centres in the fragile Himalayan environment have degraded critical ecosystems. It has been observed that rapid and unplanned urbanization has disrupted the hydrological regime of the Himalayan watersheds and has reduced groundwater recharge, reduced water availability for drinking, crop production, and sanitation, as well as decreased forests and biodiversity, increased risks of natural disasters and hazards in urban areas, as well as in peri-urban areas, and an increase in the vulnerability of mountain dwellers regarding food, water, livelihood, and health [16]. These challenges impede Himalayan cities' progress towards achieving Sustainable Development Goal 11, which emphasizes fostering inclusive, safe, resilient, and sustainable cities and settlements.

Climate change significantly impacts the WHR, causing the rise in temperatures, melting glaciers and snowpacks, changes in precipitation patterns, and increased intensity and frequency of extreme weather events. These impacts threaten the region's water resources, ecosystems, and communities. The study shows that the region is warming twice the global average rate [17]. This warming significantly impacts the region's glaciers, snowpack, precipitation patterns, and ecosystems. The study shows that temperatures in the west Himalayan (WH) region have risen by around 1.6 °C over the past century, warming winters faster than summers [18]. According to a report published in 2020 by the Ministry of Earth Sciences, Government of India, in the Hindukush Himalaya (HKH), the mean surface-air temperature increased at the rate of 0.1 degree Celsius per decade from 1901 to 2014 and from 1951 to 2014, the rate of increase has been doubled 0.2-degree Celsius. A significant warming trend has been observed at higher altitudes, where the temperature has risen to 2 °C [19]. This has led to glaciers retreating at an alarming rate, and some studies suggest that glaciers may disappear entirely within a few decades. A shift in precipitation patterns is also being observed in the region. Although total rainfall has decreased, extreme rainfall events have increased in intensity and frequency [20]. As a result, floods and landslides have become more frequent, causing widespread damage and making the Himalayan ecosystem and cities more vulnerable. Climate change acts as a force multiplier, making natural disasters such as landslides, cloudbursts, glacier lake outbursts flood (GLOF), and flash floods more disastrous in the region. The warming rate of western Himalayan cities is faster than that of the other regions [21]. For example, Dharamshala city is a western Himalayan city in Himachal Pradesh. The city is situated at an elevation of 1,745 m above the mean sea level. The city falls under the subtropical climatic zone. Figure 1 shows the temperature from 1981 to 2022. Figure 1 shows 3.0 ˚C temperature variation (minimum and maximum) from 1981 to 2022. A similar situation is happening in other western Himalayan cities, in Dehradun 0.37 ˚C, in Haridwar 0.34 ˚C, in Uttarkashi 0.51 ˚C, and in Pithoragarh 0.58 ˚C temperature increase is recorded from 1911 to 2012 [22]. The temperature change in the western Himalayan cities shows that the cities are warming over the years.

The rapid population growth and climate change simultaneously create a complex and multifaceted crisis for western Himalayan cities. Due to the rapid and unplanned growth of the western Himalayan urban centres, they are becoming increasingly

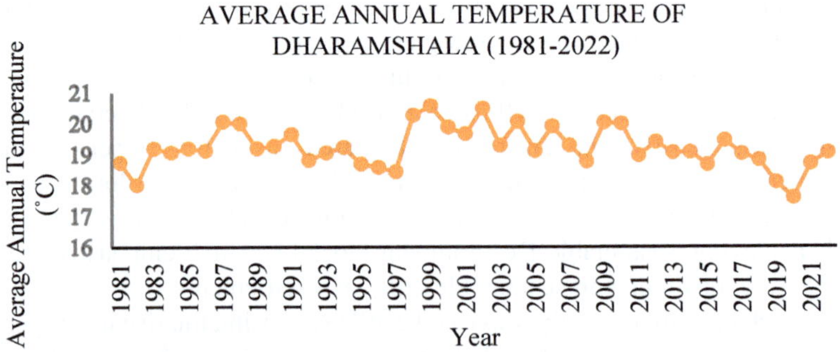

Fig. 1 Mean annual temperature trend of Dharamshala city, Himachal Pradesh [47]

vulnerable to water insecurity, natural disasters, and public health challenges. The rapid growth of urban centres leads to the degradation of mountain ecosystems, and the strain on infrastructure further complicates the problems. The combined effects of population growth and climate change pose a significant threat to urban resilience, sustainable infrastructure, and effective waste management and create hindrances in achieving SDGs-11 in the western Himalayan cities. It is essential to use a holistic and integrated tool to address these challenges. Geoinformatics can play a crucial role in monitoring and understanding the impacts of these challenges in the western Himalayan cities.

3 Geoinformatics: Concepts and Tools

Geoinformatics is the integration of geospatial data and information science to solve problems related to the Earth's environment. The word 'Geoinformatics' is made by two words: "Geo and Informatics." "Geo" means "the Earth," and "Informatics" means "the study of information processes." Geoinformatics is the combination of Earth Sciences and informatics. Thus, geoinformatics can be defined as the science that deals with the acquisition, storage, processing, presentation, and dissemination of geospatial information.

According to Burrough (2005), Geoinformatics comprises a set of tools utilized for collecting, storing, retrieving, transforming, and presenting geospatial data from the real world [23]. Similarly, Oledzki (2004) describes Geoinformatics as the fusion of science and technology dealing with the structure and characteristics of spatial data, encompassing its acquisition, classification, storage, processing, portrayal, and dissemination, along with the requisite infrastructure for optimal utilization. It is also regarded as the integration of art, science, and technology involved in the acquisition, storage, processing, production, presentation, and dissemination of geospatial information [24].

Jachimski (2001) defines geoinformatics as the science of acquiring, storing, processing, and disseminating geospatial information defined within the Earth's system [25]. It refers to the advancement of science and technology that can be used to conduct research and development in any field, such as climate change vulnerability assessment, natural hazards, land use degradation monitoring, etc. Especially for researchers, it is a powerful tool for generating and geocoding spatial data and creating an attribute database that helps them make decisions for planning according to the information available [26]. Geoinformatics is a combination of tools such as geographical information systems (GIS), remote sensing, and global positioning systems (GPS) (refer to Fig. 2) and leverages these tools to achieve its objectives.

"GIS is a computer-based system for capturing, storing, querying, analyzing, and presenting geospatial data" [27]. GIS combines software, hardware, data, and people to manage and analyze spatial data effectively. GIS plays a central and indispensable part of geoinformatics. GIS heavily relies on GIS for spatial data integration, data analysis, visualization, geodatabase management, georeferencing, environmental

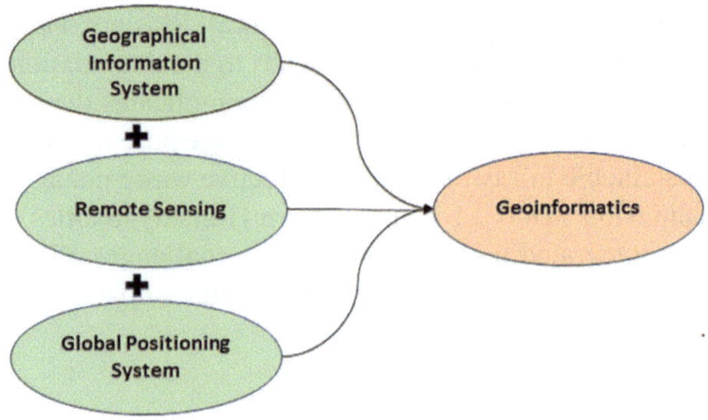

Fig. 2 Components of geoinformatics *Source* Authors

monitoring, hazard mapping and monitoring, natural resource management, urban planning and development, transportation and logistics, etc. GIS is a cornerstone of geoinformatics, enabling the effective acquisition, analysis, and management of geographic information. It is a versatile and indispensable tool for numerous applications in climate change vulnerability assessment (CCVA). It contributes to improved planning, resource management, disaster risk response, susceptible zone mapping, informed decision-making, and making people or communities more resilient to climate change effects.

Remote sensing is the basic part of the development of information technology. It has been used in various disciplines, including geography, land surveying, and Earth Sciences. It is also used for hazard monitoring, military intelligence, commercial, economic, planning, and humanitarian purposes. Remote is the art and science of obtaining information without physically touching the object [28]. Remote sensing plays a significant role in Geoinformatics by providing valuable data and information about the surface of the Earth and its various features. Geoinformatics is the science and technology of acquiring, analyzing, managing, and visualizing geoinformation, and remote sensing is the critical component of this field. Remote sensing is a critical tool that provides spatial and spectral data, which, when properly analyzed and integrated, also enhances understanding of the Earth's surface, its temporal changes, and its various features. Remote sensing is a crucial tool for regions where physical survey is not possible, especially in mountain regions. It plays a crucial role in making mountain regions more resilient to climate change by providing valuable data and information that can be used for planning and making cities more sustainable and safer. Geoinformatics addresses global challenges, including environmental conservation, urban planning and development, and disaster response, and enables decision-makers to understand and manage complex spatial problems better, leading to more informed and effective decision-making.

GPS, short for "global positioning system," is also referred to as the "Navigation System with Time and Ranging Global Positioning System (NAVSTAR)." Initially developed for military purposes, GPS has progressively found widespread applications among researchers and civilians for various purposes such as surveying, marine navigation, and location determination. It constitutes a constellation of at least 24 satellites orbiting the Earth in precisely known orbits, organized into six different orbital planes, with groups of four or more satellites in each plane.

Geoinformatics heavily relies on GPS technology for accurate positioning, data collection, mapping, navigation, surveying, land use planning, hazard monitoring, natural resource monitoring, emergency response, and more. By providing precise and real-time data, GPS enables individuals and communities to gain a better understanding of and respond effectively to the challenges posed by climate change in any given region. It plays a vital role in enhancing disaster resilience, optimizing resource management, and implementing adaptive strategies aimed at safeguarding both the mountain environment and the inhabitants of the region.

Integrating GIS, remote sensing, and GPS tools can be utilized to assess CVA in western Himalayan cities. These tools can be used to identify areas that are more vulnerable, assess the exposure and sensitivity of cities to climate change impacts, develop strategies for adaptation and mitigation, and help make cities more sustainable and resilient to climate change.

4 Geoinformatics for Climate Change Vulnerability Assessment in the Western Himalayan Cities

Vulnerability refers to the potential threat posed to people, communities, or assets by hazards. It encompasses how individuals, communities, or the environment are negatively impacted when confronted with specific hazards or disasters, including but not limited to climate change, earthquakes, floods, hurricanes, wildfires, or other natural or man-made events. There are two primary perspectives through which vulnerability is understood: the natural perspective and the social perspective.

From the natural perspective, vulnerability is viewed in terms of the magnitude and rate of climate variability to which a system is exposed, its sensitivity to these variations, and its adaptive capacity. On the other hand, the social perspective defines vulnerability as the extent to which a system is susceptible or prone to injury or damage.

Himalayan cities are particularly at risk from various types of climate change-induced hazards, such as extreme weather events, increased landslides, rockfalls, flash floods, water scarcity, and land degradation. Addressing these challenges requires the utilization of suitable tools and technologies. Climate change vulnerability assessment (CCVA) serves as a systematic evaluation of the potential threat posed by climate change to any given region or system. It entails assessing the susceptibility or vulnerability of the region or system to known climate change-induced threats, determining their severity, and recommending mitigation strategies as necessary.

Western Himalayan cities stand on the frontline of climate change impacts, confronting a barrage of extreme weather events including landslides, flash flooding, and challenges in waste management. These events inflict extensive socioeconomic losses upon the cities. Geoinformatics emerges as a pivotal tool in fortifying urban resilience by furnishing invaluable insights into climate change risks and vulnerabilities. Through the analysis of spatial and temporal patterns of extreme weather events like floods, landslides, and glacial lake outburst floods (GLOFs), geoinformatics can pinpoint areas predisposed to these hazards. This enables the implementation of proactive disaster risk reduction measures aimed at safeguarding lives and infrastructure.

Geoinformatics can be utilized from the beginning in evaluating climate change vulnerability, which can include several aspects such as data management, spatial analysis, mapping, and tracking of the impacts [29]. There are the following applications of geoinformatics and its components to make the western Himalayan cities more resilient to climate change impacts;

4.1 Geoinformatics and Urban Spatial Planning

The western Himalayan cities are experiencing unplanned growth, situated amidst ecologically sensitive mountain environments, and grappling with rapid population expansion, infrastructure development, and environmental degradation issues. Planning and developing spaces in mountainous regions pose considerable challenges due to factors such as rapid population growth, unplanned urbanization, limited land availability, susceptibility to natural disasters, and ecological sensitivity [30, 31]. The micro-level geographic variations further complicate the adoption of standardized planning norms. Each locality presents unique challenges and opportunities, necessitating the consideration of various factors in its development planning [30]. Moreover, planning becomes increasingly challenging due to conflicts between environmental conservation, economic growth, and sustainable development [31].

In this context, GIS, a vital component of geoinformatics, have emerged as indispensable tools for sustainable urban spatial planning. GIS offers a comprehensive and data-driven approach to urban development. It is a computer-based tool utilized for collecting, storing, manipulating, analyzing, and displaying spatially referenced data [32]. As a type of data-based information technology, GIS facilitates the generation of information from multiple database sets, enabling rational spatial planning procedures. Its geospatial analysis capabilities support a more informed, rational, and sustainable decision-making process grounded in scientific principles [33] [31].

By integrating various data layers, including land use, topography, infrastructure, and environmental parameters, GIS empowers urban planners to make informed decisions that strike a balance between development needs and environmental considerations. GIS can be instrumental in urban spatial planning in various ways:

4.1.1 Monitoring Urban Land Use and Land Cover Change (LULC)

Rapid urbanization and unplanned growth are swiftly altering the land use and land cover (LULC) patterns of western Himalayan cities. Monitoring these changes presents several challenges owing to the region's intricate terrain, dynamic land cover, and limited availability of ground reference data. However, the integration of GIS, remote sensing tools, and the increasing accessibility of high-resolution satellite imagery offer novel opportunities for more precise and real-time monitoring of LULC changes in the region. By harnessing remote sensing, GIS, and GPS technologies, geoinformatics provides a robust method for analyzing land use issues and a versatile tool for modeling both agricultural and urban land use. Additionally, it facilitates monitoring and mapping the evolution of urban green spaces within the cities. Through these mechanisms, geoinformatics contributes to enhancing the sustainability and resilience of Western Himalayan cities.

4.1.2 Planning of Urban Transport System

The western Himalayan cities are facing rapid urbanization and increased population density in recent years. Since the region has roughed topography and is ecologically very sensitive, a slight change in the region leads to significant challenges. Due to insufficient infrastructure, rugged topography, steep slopes, and narrow roads, there are concerns regarding road safety, air pollution, and traffic congestion in these cities. Geoinformatics, integrating geospatial data and tools, can be vital in addressing these challenges in planning sustainable transport systems. In such conditions, GIS tools can be used in mapping and analyzing existing road networks, understanding travel patterns, and monitoring and managing transportation systems in the cities [35]. It can facilitate the development of sustainable, efficient, and environmentally friendly transportation systems that cater to the unique needs of the region and make cities more resilient.

4.1.3 Distribution of Essential Services

In the western Himalayas, a vast and diverse landscape cradle to many cities, climate change poses a significant threat. Due to its unique topography, characterized by steep mountains and glaciers, this region is particularly at risk from rising temperatures, shifts in precipitation patterns, and melting ice. In response to these changes, extreme weather events are becoming more frequent, such as flash floods and landslides, which disrupt essential services and threaten millions of lives and livelihoods.

In this context, it is crucial to plan and deliver essential services to build resilience to climate change. Essential services, such as water supply, sanitation, healthcare, and education, are critical for human well-being and economic development. Integrating GIS, remote sensing, and GPS can collect, manage, and analyze spatial data, providing real-time information about the region's geography, infrastructural change, LULC change, and population distribution.

This spatial information serves as a foundation for informed decision-making regarding resource allocation and the design of service delivery networks. For instance, GIS aids in the distribution and monitoring of water facility centers within cities, enabling optimal location selection based on population distribution. Similarly, as urban populations grow, the demand for educational centers and other essential services escalates. In this context, GIS plays a pivotal role in identifying and monitoring the distribution of these services in accordance with population density [36].

4.2 *Geoinformatics for Natural Hazards Monitoring*

The Western Himalayan region boasts a rich and diverse ecosystem; however, it is also among the most susceptible to natural hazards. Under the climate change scenario, the frequency of natural hazards has increased [4], and the western Himalayan cities are facing such events. The impacts of climate change are exacerbating landslides, flash floods, rockfalls, etc., in the cities and putting millions of people at risk.

Geoinformatics, characterized by the integration of geospatial tools and spatial data analysis techniques, emerges as a potent instrument for mapping and monitoring natural hazards across the globe. Through the utilization of real-time satellite imagery and GIS tools, geoinformatics provides a comprehensive understanding of the spatial distribution, patterns and trends, and drivers of natural hazards. This facilitates proactive risk assessment, mitigation planning, and formulation of disaster response strategies, ultimately bolstering the resilience to climate change.

In the context of western Himalayan cities, geoinformatics can be harnessed in the following ways to mitigate and minimize the risk of climate change-induced natural hazards:

4.2.1 Geoinformatics as an Early Warning System

Early warning systems leveraging GIS and real-time remote sensing data offer critical information regarding natural disasters and emergency events, empowering individuals to take proactive measures to safeguard lives and infrastructure. For instance, in western Himalayan cities, landslides reign as the most prevalent natural hazard, often stemming from extensive deforestation and slope modifications for infrastructure development. Real-time satellite imagery and GIS tools facilitate the monitoring of slope movements. By identifying areas prone to landslides based on slope instability, early alerts can be disseminated to residents and relevant authorities, enhancing preparedness and response efforts.

Moreover, such systems can serve as early warning mechanisms for flood and earthquake monitoring in western Himalayan cities, offering vital time for mitigation and evacuation measures to be implemented.

4.2.2 Geoinformatics and Urban Flash Flood Monitoring

Flash flooding is rapid inundation in low-lying areas caused by heavy rainfall in a short time. It can occur within minutes or hours. Short, intense rainfall may cause flash floods due to rapid discharges and sudden increases in water levels in rivers and streams. During the monsoon season, the Indian Himalayan region (IHR) experiences flash flooding yearly. The western Himalayan cities are vulnerable to the devastating impacts of flash floods due to rapid urbanization, dense human settlements, rugged topography, glacial lake outbursts, and inadequate drainage systems. In these cities, GIS and remote sensing can be used in the following ways to monitor and map the flash floods to make cities more sustainable and climate change resilient [37];

a. Mapping of flood-prone areas.
b. Hydrological assessment of rivers and channels.
c. Real-time monitoring of rainfall and stream flow.
d. development of an early warning system by using remote sensing and GIS tools.
e. Mapping and monitoring of urban land use change patterns.
f. Real-time mapping and monitoring of glacial lakes in higher elevations.

4.2.3 Geoinformatics and Landslide Hazards Monitoring

Landslides represent a significant natural hazard in the western Himalayan cities, resulting in substantial damage to people, infrastructure, and the economy. Recent studies indicate a surge in landslide frequency attributed to climate change-induced factors, such as increased intensity, duration, and pattern of rainfall [38]. Given the variability in rainfall patterns across the Himalayan region, the majority of landslides are triggered by rainfall [39]. However, assessing landslide hazards poses a formidable challenge, traditionally requiring laborious and time-consuming processes due to manual data handling and processing [40].

In recent years, the integration of GIS and remote-sensing techniques has emerged as an indispensable tool for landslide hazard assessment. GIS and remote-sensing techniques can be leveraged for landslide hazard assessment in the following ways:

4.2.4 Preparation of Landslide Inventories

Landslide inventory is a comprehensive database of past landslide events in an area. A landslide inventory includes information regarding the location, date, type, and size of each landslide. Landslide inventories are essential for understanding landslide risks, susceptibility, and frequency of failure and help in mitigation strategies. GIs and remote sensing play a crucial role in preparing landslide inventories of an area. As we know, mountainous areas are not easily accessible, and the inaccessibility creates hurdles in preparing landslide inventory in the western Himalayan cities. In such conditions, the preparation of landslide inventories is very costly and time-consuming. In this context, remote sensing and GIS techniques provide real-time data on inaccessible mountain regions, reducing time and cost.

4.2.5 Preparation of Landslide Susceptibility Map

Landslide susceptibility refers to the probability of a slope failure happening in a particular region based on the local characteristics of the terrain [41, 42]. It predicts the potential location of landslide occurrences [43]. Landslide susceptibility mapping involves identifying areas that are more prone to experiencing landslides based on the characteristics of factors influencing landslide occurrences.

Rapid urbanization and unplanned constructions in the western Himalayan cities have led to an increase in landslide frequency over the years. In this context, GIS and remote sensing play a crucial role in the real-time monitoring and mapping of susceptible zones in these cities. The integration of GIS, remote sensing, and statistical or machine learning techniques enables precise and accurate mapping of landslide susceptibility. This significantly contributes to disaster preparedness and risk reduction, thereby enhancing the resilience of cities to climate change.

4.2.6 Geoinformatics for Landslide Hazard Monitoring and Disaster Response

Geoinformatics plays a significant role in slope stability monitoring and post-landslide response. It provides detailed information about landslide-prone areas, real-time slope stability monitoring, and support for preparedness and response planning. Geoinformatics is a power tool that can be used to monitor landslide activity in the western Himalayan cities for early warning and mitigation processes after slide. Remote sensing and GIS technologies can be utilized to quickly assess the extent of damage and map the area that can be used to guide relief workers and emergency responders.

4.3 Geoinformatics for Municipal Solid Waste Management (MSWM)

Rapid urbanization and population growth in the western Himalayan cities have resulted in a significant surge in municipal solid waste (MSW) generation. As urban centres expand, the production of MSW steadily escalates over time [44,

45]. However, the challenging terrain, harsh climate conditions, limited land availability for waste disposal, remoteness, and poor accessibility render the collection of MSW a daunting task in western Himalayan cities. Furthermore, the impacts of climate change further compound the challenges of municipal solid waste management (MSWM) by increasing the frequency of extreme weather events, such as flash floods and landslides. These events can disrupt MSW collection and disposal services, posing additional threats to human health and the environment.

In this context, the integration of GIS, remote sensing, and global navigation satellite systems (GNSS) offers a novel approach to improve MSWM in the western Himalayan cities in the face of climate change. By furnishing geospatial data and analytical tools, geoinformatics can serve the following purposes:

4.3.1 Waste Collection and Transportation

Remote sensing and GIS techniques offer the means to track waste collection vehicles, optimize collection routes, and monitor real-time waste collection performance. This can improve efficiency, reduce fuel consumption, and minimize the environmental impact of waste transportation.

4.3.2 Waste Disposal and Landfill Monitoring

Remote sensing and GIS technology offer valuable tools for monitoring landfill sites, enabling the detection of potential environmental impacts like leachate contamination and groundwater pollution. These technologies facilitate the assessment of landfill capacity, aid in planning for future expansion, and enable the monitoring of compliance with environmental regulations.

Geoinformatics can lead to more efficient, sustainable, and climate-resilient MSW management practices, minimizing the environmental impact of waste, protecting public health, and helping make the western Himalayan cities more resilient to climate change.

5 Limitations of Geoinformatics

Although geoinformatics is an essential tool for climate change vulnerability (CCV) and informing adaptation strategies, it also has several limitations. The limitations of geoinformatics in CCVA for the western Himalayan cities are as follows:

5.1 Data Availability

The main limitation of geoinformatics in climate change vulnerability assessment lies in the availability and quality of geospatial data. In mountainous regions, such as the Himalayas, acquiring accurate and precise data can be challenging due to the complex topography, limited accessibility, and paucity of ground-truthing observations. These factors can lead to uncertainties in vulnerability assessments, potentially misrepresenting the actual exposure and sensitivity of cities to climate change impacts.

5.2 Scale and Resolution

Scale and resolution of remotely sensed data are very important because Geoinformatics tools often operate at specific scales, which may not adequately capture the heterogeneity and complexities of urban environments. For instance, while satellite imagery provides broad-scale insights into land use patterns and urban expansion, it may overlook finer-grained details such as localized infrastructure vulnerability or the distribution of vulnerable populations within the city. The scale mismatching can lead to an incomplete understanding of vulnerability and hinder the development of targeted adaptation measures.

5.3 Lack of Ground Truthing

Ground truthing is a very important step in remote sensing based. It is the process of collecting data on the ground to verify the accuracy of remotely sensed data. It is essential because remotely sensed data can be affected by various factors, such as atmospheric conditions, sensor calibration, and image processing algorithms. Without ground truthing, it can be challenging to ensure the accuracy of remotely sensed data.

5.4 Community Engagement and Local Knowledge

Effective climate change vulnerability assessments require the participation of local knowledge and perspectives. While geoinformatics provides valuable spatial insights, it is crucial to engage with communities in the Western Himalayas to understand their lived experiences of climate change and their priorities for adaptation. This participatory approach can lead to more nuanced and actionable vulnerability assessments that reflect the realities of these communities.

Table 1 Urban population growth in Uttarakhand from 1951 to 2011[46]

Year	Urban population	Urban population to total population (%)	Decadal growth (%)	No. of towns
1951	398,296	13.52	32.11	34
1961	494,966	13.71	16.53	32
1971	734,276	16.34	32.59	40
1981	1,061,821	15.54	30.85	65
1991	1,579,295	22.40	32.77	76
2001	2,184,849	25.74	27.72	86
2011	3,026,203	30.00	27.80	115

5.5 Requirement of Specialized Software and Hardware

The requirement of notable software and hardware also limits geoinformatics. These software and hardware can cost thousands of dollars and are very expensive. These software and hardware requirements may make geoinformatics less accessible, especially in areas with limited technological infrastructure.

6 Conclusion

The western Himalayan cities are highly vulnerable to the effects of climate change, as evidenced by an increase in the frequency and intensity of extreme weather events such as flooding, landslides, and rockfalls. These impacts are projected to exacerbate in the future due to changing climatic conditions, posing a significant threat to both people and infrastructures. Geoinformatics holds substantial potential in climate change vulnerability assessment (CCVA) by furnishing critical insights into the determinants of vulnerabilities within mountainous regions. It aids in the planning and management of urban centres in mountainous areas amidst climate change challenges. Geoinformatics facilitates the mapping and quantification of the frequency and intensity of extreme weather events, as well as the exposure and sensitivity to natural hazards. It proves instrumental in identifying susceptible zones to natural hazards, aiding in disaster risk response, and facilitating the development of climate change adaptation plans and strategies. Through modelling the potential impacts of climate change on urban areas, geoinformatics can pinpoint more vulnerable areas and devise targeted interventions to mitigate vulnerability, thereby enhancing the resilience of mountain cities to climate change.

References

1. Bookhagen B, Burbank DW (2006) Topography, relief, and TRMM-derived rainfall variations along the Himalaya. Geophysical Research Letters, 33(8)
2. Rasul G (2014) Food, water, and energy security in South Asia: A nexus perspective from the Hindu Kush Himalayan region {star, open}. Environ Sci Policy 39:35–48
3. Sinha, A. K., & Upadhyay, R. (1995). Himalaya: Geological aspect. Journal of Palaeosciences, 44, 9–28. https://doi.org/10.54991/jop.1995.1198
4. Das L, Meher JK (2019) Drivers of climate over the Western Himalayan region of India: A review. Earth Sci Rev 198(July):102935
5. Urban Himalaya.: Research Urban Himalaya. https://urbanhimalaya.yale.edu/research, (2023).
6. Cruz RV, Harasawa H, Lal M, Wu S, Anokhin Y, et al.: Asia. In: Climate Change 2007: Impacts, Adaptation and Vulnerability. Contribution of Working Group II to the Fourth Assessment Report of the Intergovernmental Panel on Climate Change, ML Parry, OF Canziani, JP Palutikof, PJ Linden VD, and Hanson, C.E. (Eds.) Cambridge University, (2007).
7. Intergovernmental Panel on Climate Change (IPCC) (2021) Mountains. in climate change 2022–Impacts, Adaptation and Vulnerability. https://doi.org/10.1017/9781009325844.022
8. Kroël-Dulay G, Ransijn J, Schmidt IK, Beier C, De Angelis P, De Dato G, Dukes JS, Emmett B, Estiarte M, Garadnai J, Kongstad J, Kovács-Láng E, Larsen KS, Liberati D, Ogaya R, Riis-Nielsen T, Smith AR, Sowerby A, Tietema A, Penuelas J (2015) Increased sensitivity to climate change in disturbed ecosystems. Nat Commun 6:1–7
9. Singh SP, Singh V, Skutsch M (2010) Rapid warming in the Himalayas: Ecosystem response and developmental options. Climate Dev 2:1–13
10. Houet T, Loveland TR, Hubert-Moy L et al (2010) Exploring subtle land use and land cover changes: a framework for future landscape studies. Landscape Ecol 25:249–266
11. Schilling Janpeter, FreierKorbinian P, HertigeElke, Scheffran Jurgen.: Climate change, vulnerability and adaptation in North Africa with focus on Morocco. Agriculture, Ecosystems and Environment 156:12–26, (2012).
12. Messerli, B. & J.D Ives (eds).: Mountains of the World. A Global Priority. - London, New York: Parthenon Publishing, (1997).
13. 'United States Agency International Development', pp. 1–183, (2016).
14. https://censusindia.gov.in/census.website/ (2011).
15. Diksha and Kumar, A.: 'Analysing urban sprawl and land consumption patterns in major capital cities in the Himalayan region using geoinformatics,' Applied Geography, 89(April), pp. 112–123, (2017).
16. Tiwari P, Tiwari A, Joshi B (2018) Urban Growth in Himalaya: Understanding the Process and Options for Sustainable Development. Journal of Urban and Regional Studies on Contemporary India 4(2):15–27
17. Pepin NC, Bradley S, Diaz HF, Baraer M, Caceres EB, Forsythe N et al (2015) Elevation-dependent warming in mountain regions of the world. Nat Clim Chang 5(5):424–430
18. Bhutiyani MR, Kale VS, Pawar NJ (2007) Long-term trends in maximum, minimum and mean annual air temperatures across the Northwestern Himalaya during the twentieth century. Clim Change 85(1–2):159–177
19. Shekhar MS et al (2010) Climate-change studies in the western Himalaya. Ann Glaciol 51(54):105–112
20. Wester, P. et al. The Hindu Kush Himalaya Assessment: Mountains, Climate Change, Sustainability and People, The Hindu Kush Himalaya Assessment: Mountains, Climate Change, Sustainability and People, (2019).
21. Shrestha, U.B., Gautam, S. and Bawa, K.S.: 'Widespread climate change in the Himalayas and associated changes in local ecosystems.', PloS one, 7(5), pp. 1–10, (2012).
22. Mishra, A.: Changing Temperature and Rainfall Patterns of Uttarakhand. International Journal of Environmental Sciences & Natural Resources, 7(4), (2017).
23. Burrough P.A. and Mc Donnell R.A.: Principles of Geographical information systems, Oxford University Press. UK. (2005).

24. Jan Romuald Olędzki: Geoinformatics – An Integrated Spatial Research Tool. Miscellanea Geographica Warszawa Vol. 11, (2004).
25. Jachimski, J. (2001) Panel discussion devoted to matters of terminology in geoinformatics; in Polish, Geoinformatyka Polonica, Kraków, No. 3, pp.57–79. Crossreferenced from Oledzki, J. R. (2004)
26. Mankari M.P, Kodge B.G, Kulkarni M.J., N. A. U.: Fundamentals of geoinformatics and its applications in geography. Geoscience Research, 1(1), 01–06, (2010).
27. Kang-Tsung Chang.: Introduction to Geographic Information System. McGraw-Hill Education, 2 Penn Plaza Newyork, NY 10121, (2019).
28. Lillesand, Kiffer &Chipman.: Remote Sensing and Image Interpretation. John Willey & Sons, USA, (2015).
29. Kienberger, S., Lang, S. & Zeil, P.: Spatial vulnerability units – expert-based spatial modelling of socioeconomic vulnerability in the Salzach catchment. Austria. In: Natural Hazards Earth System Sciences, 9, 767–778, (2009).
30. Maitra, A. K.: "Development of Hill Capital: Shimla-2035". Proceedings of 52nd National Town and Country Planning Conference on Development of Hill Capitals: Shimla Vision 2025, pp. 9–15, (2003).
31. Kapoor N, Jain M, Bansal VK (2019) GIS-based framework for local spatial planning in Hill Areas. International Review for Spatial Planning and Sustainable Development 7(4):113–130
32. Burrough, P. A., and Rachael A. McDonnell.: Principles of Geographical Information Systems. New York: Oxford University Press, (1998).
33. Ervin, S.: "A System for Geodesign". Proceedings of Digital Landscape Architecture, Anhalt University of Applied Sciences, pp. 145–154, (2011).
34. Abdel-Rahman A. Mustafa and Osama E. Negim.: Utilizing of Geoinformatics for Mapping Land Use/Land Cover Changes in Sohag, Egypt. American Journal of Environmental Engineering and Science, 3(1): 33–42, (2016).
35. Pouucos Prastacos.: Integrating GIS Technology in Urban Transportation Planning and Modeling. Transportation Research Record 1305.
36. Pushpalata Shah.: IGiS For Managing Cities Smartly - Urban Geoinformatics. The International Archives of the Photogrammetry, Remote Sensing and Spatial Information Sciences, Volume XLVI-4/W5–2021
37. Ding L, Ma L, Li L, Liu C, Li N, Yang Z, Yao Y, Lu H (2021) A Survey of Remote Sensing and Geographic Information System Applications for Flash Floods. Remote Sens 13:1818
38. Kumari, S., Chauhan, A. and Shankar, V.: 'Assessment of climate change implications on landslides in mid and high hills of Himachal Pradesh, India', Arabian Journal of Geosciences, 14(14), (2021).
39. Dikshit A, Sarkar R, Pradhan B, Segoni S, Alamri AM (2020) Rainfall induced landslide studies in Indian Himalayan region: A critical review. Applied Sciences (Switzerland) 10(7):1–24
40. Huabin W et al (2005) GIS-based landslide hazard assessment: An overview. Prog Phys Geogr 29(4):548–567
41. Brabb, E.E.: Innovative approaches to landslide hazard mapping. In Proceedings of 4th International Symposium on Landslides, Toronto, 307–24, (1984).
42. Reichenbach, P. et al.: 'A review of statistically-based landslide susceptibility models', Earth-Science Reviews, 180(November 2017), pp. 60–91, (2018).
43. Guzzetti F, Peruccacci S, Rossi M, Stark C (2007) Rainfall thresholds for the initiation of landslides in central and southern Europe. Meteorol Atmos Phys 98:239–267
44. Singh A (2019) Remote sensing and GIS applications for municipal waste management. J Environ Manage 243(May):22–29
45. Raviv O, Broitman D, Ayalon O, Kan I (2018) A regional optimisation model for waste-to-energy generation using agricultural vegetative residuals. Waste Manag 73:546–555
46. Pant BR, Chand R (2020) Characteristics of Urban Centres and Urbanization in Uttarakhand. Journal of Urban and Regional Studies on Contemporary India. 6(2):1–20
47. https://power.larc.nasa.gov/data-access-viewer/

Assessing and Mapping Tourism Vulnerability Using Remote Sensing and Spatial Analysis Techniques: A Case of Shimla District in Himachal Pradesh, India

Keerti Manisha, Inderpal Singh, and Minakshi Jain

1 Introduction

Globally, tourism sector witnessed a four percent increase in 2021 amounting to 15 million more international tourist as compared to 2020. The sector is expected to grow further in coming years which could impose serious economic, geopolitical, and environmental concerns [1]. Tourism sector generates revenue and employment [2] along with the engagement of several stakeholders such as locals, visitors, planners, service providers, businessmen, entrepreneurs, infrastructure experts, and engineers [3]. Numerous studies have emphasized on the importance of mapping tourism vulnerability to frame effective strategies for risk reduction, resilient urban and rural areas, and sustainable communities [4]. Tourism vulnerability is influenced by factors such as; natural hazards, climate change, rise in temperature, economic recession, fluctuation in rainfall patterns, and political disturbances [5]. The climate change is imposing severe challenges on tourism and other sectors. Therefore, it is essential to conduct spatial assessments to identify vulnerable areas [6]. Numerous researchers have applied analytic hierarchy process (AHP) tool for mapping the tourism vulnerability to several internal and external factors, which can negatively impact attractiveness and sustainability [7]. The Indian Himalayan region (IHR) is extremely

K. Manisha (✉)
Department of Architecture, National Institute of Technology Tiruchirappalli, Tiruchirappalli, Tamil Nadu, India
e-mail: ar.keertimanisha@gmail.com; keerti@nitt.edu

I. Singh · M. Jain
Department of Architecture, National Institute of Technology Hamirpur, Hamirpur, Himachal Pradesh, India
e-mail: ipsingh05@gmail.com; ipsingh@nith.ac.in

M. Jain
e-mail: jainminakshi@gmail.com; minakshi@nith.ac.in

V. Chettry et al. (eds.), *Geoinformatics for Hilly Regions*, Advances in 21st Century Human Settlements, https://doi.org/10.1007/978-981-95-0156-4_11

vulnerable to climate change causing adverse impacts such as glacier shrinking, forest degradation, water scarcity, and reduced rainfall [8, 9]. These impacts are anticipated to get worse in coming times due to increased tourist arrivals. In this context, this study attempted to examine the tourism vulnerability in Shimla district in Himachal Pradesh state using remote sensing and geographic information system (GIS). The study has following objectives; a) identifying indicators for tourism vulnerability assessment, b) a pair-wise comparison matrix, c) cumulative tourism vulnerability mapping, d) composite mapping of tourism vulnerability, and e) categorize areas under low, medium, or high degree of tourism vulnerability. The tourism vulnerability assessment allows the policymakers to develop targeted interventions to mitigate the risks imposed by fast paced development and ensure benefits to local communities from tourism. The district could promote community-based tourism, investing in capacity building, and raising awareness, and strengthen environmental protection measures in the study area. The novel contribution of this research is the comprehensive technique to select appropriate indicators for assessing and mapping tourism vulnerability in hilly region facing multiple susceptibilities. The rest of the paper is structured as follows: literature review and theoretical foundation, data and research methodology, study area, results and discussions, and conclusions.

2 Literature Review and Theoretical Foundation

The literature on vulnerability has grown manifolds in recent years and covers ecological "disaster vulnerability, ecosystem vulnerability, and climate change" [10]. The vulnerability is assessed in terms of "exposure, sensitivity, and adaptive capacity," however, there is a serious gap in collective assessment [11]. Tourism has been studied with respect to its impact on the livelihood [12], economic sustainability [8, 13, 14], impact on coastal and beach tourism [15], vulnerability of coastal destinations [16], ecological vulnerability of tourism spots [17], vulnerability and resilience of tourism sector [18], over tourism in Indian cities [19] and vulnerability and resilience in the protected areas [20]. AHP as a decision-making tool evaluates multiple criteria using hierarchical process which essentially maps the vulnerability through the pairwise comparison matrix [15]. Although tourism and climate are linked together in the study area but there are no studies explicitly on tourism vulnerability. Moreover, the scientific assessment of tourism vulnerability at district level has not been the focus of researchers in the study area [21]. This study bridges the existing gap in the literature on district level tourism vulnerability in hill areas.

3 Study Area

Shimla district is situated in Himachal Pradesh and has a geographical area of 5136 sq km. The district lies between north latitude 30°43′00" and 30°45′48" and east longitude 76°59′22" and 78°18′40". Shimla, the capital, is in Shimla district adjoining Sirmaur, Kinnaur, Bilaspur, Solan, mandi, and Kullu districts. The district has one municipal corporation, which was established in 1969 under the Himachal Pradesh Act 1968 of development and regulation. The district is connected by air, rail, and road network, refer Fig. 1.

"The population of the district is 8,14,010 as per Census 2011, of which 4,25,039 (52 %) are males and 3,88,971 (48 %) females." The climate of the district is "sub-tropical in the valleys and tends to be temperate at the hilltops." The "average annual rainfall in the district is about 999.64 mm." The "district presents an intricate mosaic of high mountain ranges, hills, and narrow, deep valleys with altitudes ranging from 1000 to 3000 m above MSL." The major drainage network is from the basins of the river Sutlej. There are major seasons such as "winter (October–March), summer (April–June), and monsoon (July–September)."

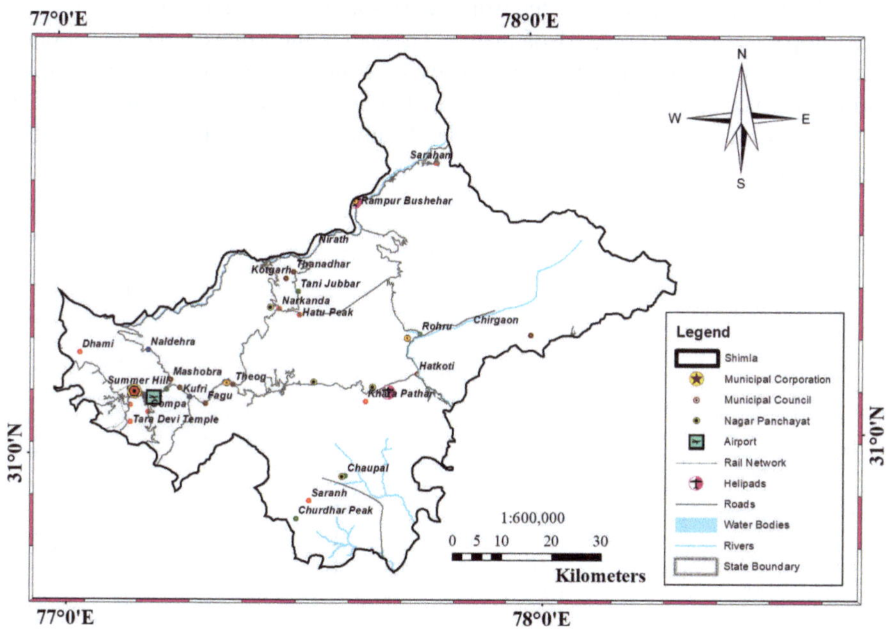

Fig. 1 Study area (*Source* Author)

4 Data and Research Methodology

Figure 2 demonstrates the research methodology adopted. This study collected data using online sources, primarily the official webpage of the concerned departments, refer Table 3. The research framework revolves around the three criteria of vulnerability; "exposure, sensitivity, and adaptive capacity" [22]. A total of 50 questionnaires were sent to urban and environmental planners (15), professors and researchers (13), officials at the state tourism department (7), architects (5), experts from tourism and travel management (5), social science (3) and psychology (2) at reputed university through electronic mail to seek a condensed list of indicators specific to the study. The experts categorised the indicators into "exposure, sensitivity, and adaptive capacity" to the best of their knowledge and proficiency. A total of 18 indicators were identified to assess and map the tourism vulnerability; 1) exposure are disaster, NDVI, elevation, slope, tourist demand, natural hazard, and LULC; 2) sensitivity includes, marginal workers, population density, water availability, GSDP, forest area, and net sown area; 3) adaptive capacity includes tourism destinations, telecommunication, electricity (hydro), medical facilities, road density, and accommodation (see Fig. 3).

The weightage of 18 indicators was calculated "using the row geometric mean method." The information collected from expert survey was fed into Microsoft Excel to develop the pairwise comparison matrix. Thereafter, the values in the consolidated pairwise comparison matrix was divided by its column total value to generate the normalized pairwise matrix using the Eq. 2. Here, γ is the normalised indicator, C is the comparison value, n is the number of indicators, and $\sum_{i=1}^{n} C$ is the summation of comparison values.

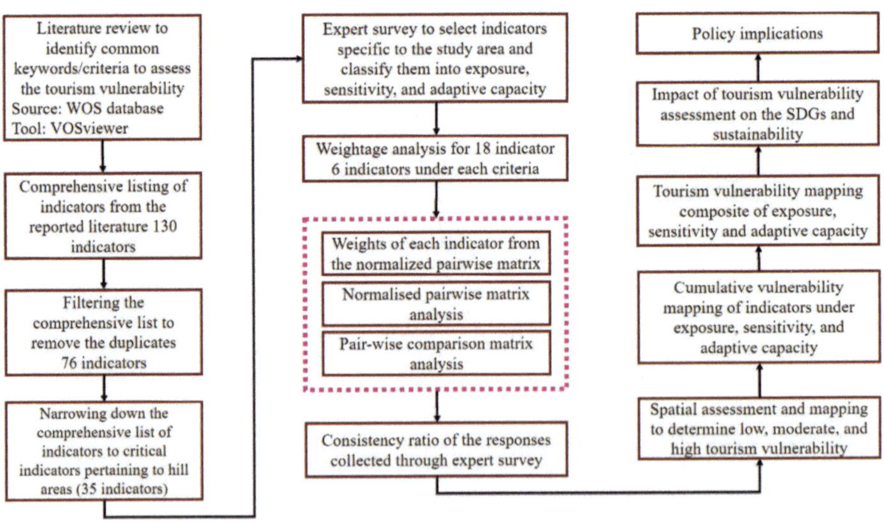

Fig. 2 Research methodology adopted in the study (*Source* Author)

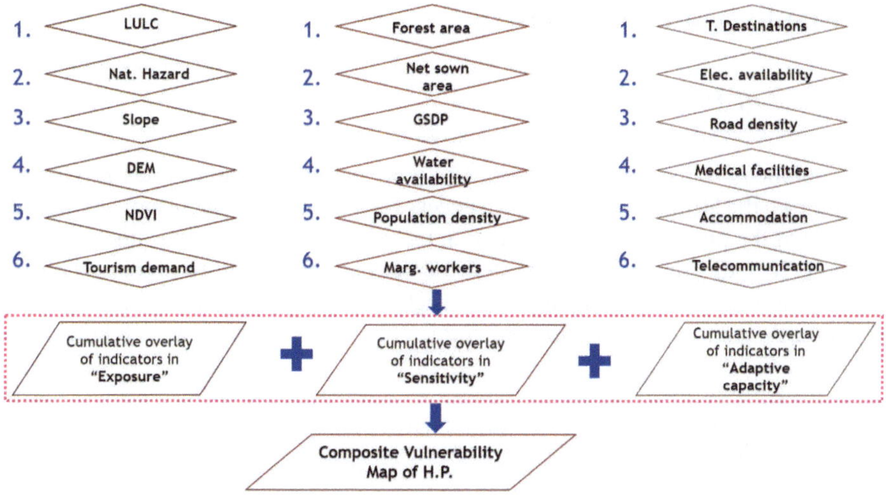

Fig. 3 Tourism vulnerability mapping approach (*Source* Author)

$$\gamma = \frac{C}{\sum_{i=1}^{n} C} \tag{2}$$

In Eq. 3, ω represents the weight of the indicator, n is the number of indicators, and γ is the normalized indicator. Lastly, the consistency among the expert responses was confirmed using Eq. (4). Saaty (2001) indicated a single numerical index consistency ratio (CR). In Eq. (4), "CI represents consistency index and RI is the random index."

$$CR = \frac{CI}{RI} \tag{4}$$

The standard value of RI depends on the order of the matrix developed by [23] is 1.24 for the indicators, refer Table 4. The consistency ratio was calculated using Eq. (5),

$$CI = \frac{\Delta - n}{n - 1} \tag{5}$$

The Δ is the consistency index (CI) which is calculated using Eq. (6).

$$\Delta = \sum_{i=1}^{n} CV \tag{6}$$

5 Results and Discussions

5.1 List of Indicators

A comprehensive list of 35 indicators was prepared in the field of tourism vulnerability. From the 50 questionnaires, the success rate of receiving the response was 86% because of the purposive sample selection technique. From the first round of Delphi method, 18 indicators were shortlisted for the data collection to prepare thematic maps. Cumulative overlay analysis for each criterion was done in ArcGIS 10.4. From the round two of the Delphi, the indicators were assigned the weightage.

5.2 Weightage Analysis

The consolidated pairwise comparison matrix of the 18 indicators was calculated using the row geometric mean method. The results are shown in Table 1.

5.3 Normalized Pairwise Matrix Analysis

Table 2 exhibits the criteria-wise normalized pairwise matrix of the indicators.

5.4 Weights of 18 Indicators from the Normalized Pairwise Matrix

Table 3 exhibits the criteria along with the selected indicators, unit, source, year of data procurement, weightage percentage, class, class range (CR), and ranks (R).

5.5 Consistency Ratio of the Responses Collected Through Expert Survey

The calculated CI was 0.04 (exposure), 0.04 (Adaptive capacity), and 0.02 (sensitivity). The value of CR in the pairwise comparisons was less than 10%, indicating a realistic consistency level, refer Table 4. The CR values confirm that the weight of the indicators is realistic and suitable for the tourism vulnerability mapping in the study area.

Table 1 Consolidated pairwise comparison matrix of the indicators

Exposure (E1 = NDVI; E2 = DEM; E3 = Slope; E4 = Tourist Demand; E5 = Natural Hazard; E6 = LULC)						
Indicator	E1	E2	E3	E4	E5	E6
E1	1.00	2.31	2.08	4.51	3.89	3.97
E2	0.43	1.00	1.56	2.76	3.00	4.06
E3	0.48	0.64	1.00	2.35	3.16	4.16
E4	0.22	0.36	0.43	1.00	1.78	2.76
E5	0.26	0.33	0.32	0.56	1.00	2.32
E6	0.25	0.25	0.24	0.36	0.43	1.00
Total	2.64	4.90	5.62	11.55	13.27	18.27
Adaptive capacity (AC1 = Tourism destination; AC2 = Telecommunication; AC3 = Electricity; AC4 = Medical facilities; AC5 = Road density, AC6 = Accommodation)						
Indicator	AC1	AC2	AC3	AC4	AC5	AC6
AC1	1	0.53	0.52	3.75	5.00	4.96
AC2	1.90	1	1.12	2.88	5.05	4.69
AC3	1.94	0.90	1	4.07	6.12	7.47
AC4	0.27	0.35	0.25	1	2.73	2.82
AC5	0.20	0.20	0.16	0.37	1	1.88
AC6	0.20	0.21	0.13	0.36	0.53	1
Total	5.51	3.18	3.18	12.42	20.44	22.82
Sensitivity (S1 = Marginal Workers; S2 = Population Density; S3 = Water availability; S4 = GSDP; S5 = Forest area; S6 = Net sown area)						
Indicator	S1	S2	S3	S4	S5	S6
S1	1	0.61	1.93	3.03	3.03	3.35
S2	1.64	1	3.37	2.83	4.24	3.27
S3	0.52	0.30	1	1.01	1.63	1.37
S4	0.33	0.35	0.99	1	1.74	2.13
S5	0.33	0.24	0.61	0.58	1	1.13
S6	0.30	0.31	0.73	0.47	0.88	1
Total	4.12	2.80	8.64	8.92	12.52	12.25

6 Spatial Assessment and Mapping of Tourism Vulnerable Areas

The analysis of expert opinion revealed electricity availability (32%), telecommunication (28%), and tourism destinations (22%) affect adaptive capacity. NDVI (36%), elevation (22%), and slope (19%) increase the degree of exposure. Population density (34%), marginal workers (26%), GSDP (12%), and water supply (12%) increase the level of sensitivity of the region. Each indicator was classified to map the tourism

Table 2 Normalized pairwise matrix of the indicators

Exposure (E1 = NDVI; E2 = DEM; E3 = Slope; E4 = Tourist Demand; E5 = Natural Hazard; E6 = LULC)						
Indicator	E1	E2	E3	E4	E5	E6
E1	0.38	0.47	0.37	0.39	0.29	0.22
E2	0.16	0.20	0.28	0.24	0.23	0.22
E3	0.18	0.13	0.18	0.20	0.24	0.23
E4	0.08	0.07	0.08	0.09	0.13	0.15
E5	0.10	0.07	0.06	0.05	0.08	0.13
E6	0.10	0.05	0.04	0.03	0.03	0.05
Adaptive capacity (AC1 = Tourism destination; AC2 = Telecommunication; AC3 = Electricity; AC4 = Medical facilities; AC5 = Road density, AC6 = Accommodation)						
Indicator	AC1	AC2	AC3	AC4	AC5	AC6
AC1	0.18	0.17	0.16	0.30	0.24	0.22
AC2	0.35	0.31	0.35	0.23	0.25	0.21
AC3	0.35	0.28	0.31	0.33	0.30	0.33
AC4	0.05	0.11	0.08	0.08	0.13	0.12
AC5	0.04	0.06	0.05	0.03	0.05	0.08
AC6	0.04	0.07	0.04	0.03	0.03	0.04
Sensitivity (S1 = Marginal Workers; S2 = Population Density; S3 = Water availability; S4 = GSDP; S5 = Forest area; S6 = Net sown area)						
Indicator	S1	S2	S3	S4	S5	S6
S1	0.24	0.22	0.22	0.34	0.24	0.27
S2	0.40	0.36	0.39	0.32	0.34	0.27
S3	0.13	0.11	0.12	0.11	0.13	0.11
S4	0.08	0.13	0.11	0.11	0.14	0.17
S5	0.08	0.08	0.07	0.06	0.08	0.09
S6	0.07	0.11	0.08	0.05	0.07	0.08

vulnerability in the district. The class range and ranking are elaborated in Table 3, for example, DEM has five classes; 1,2,3,4, and 5. The thematic maps of indicators under exposure, sensitivity, and adaptive capacity criteria are shown in Figs. 4, 5, 6, respectively. All six thematic maps under the selected criteria were overlayed to create cumulative maps based on their influence (see Fig. 7). In the cumulative maps it was noted that the upper belt of the district is exposed to very high tourism vulnerability and has low to moderate adaptive capacity. The degree exposure and sensitivity eventually reduce with enhanced adaptive capacity. Finally, three cumulative maps were overlaid to generate the composite tourism vulnerability map for the district. Primarily the district is moderately vulnerable to tourism growth in 1501 sq. km area, whereas 614 sq. km area is highly vulnerable (see Table 5).

Table 3 Dataset description, source, and weightage to each indicator

S.no	C	Indicators	Unit	Data source	Year	W(%)	Class	CR	R
1	AC	Medical facilities	Numbers	Hospital List- Mukhya Mantri Himachal Health Care Scheme-HIMCARE	2022	9	6–10	VH	5
							11–16	H	4
							17–30	M	3
							31–40	L	2
							41–54	VL	1
2	AC	Road density	Kilometer polyline	Department of Tourism and Civil Aviation (DoTCA)	2020	5	1	VH	5
							1–2	H	4
							2–4	M	3
							4–7	L	2
							7–9	VL	1
3	AC	Accommodation	Numbers	Department of Tourism and Civil Aviation (DoTCA)	2022	4	1070–7104	VH	5
							7104–14,207	H	4
							14,207–21,311	M	3
							21,311–28,414	L	2
							28,414–35,518	VL	1
4	AC	Tourism destination	Numbers	Department of Tourism and Civil Aviation (DoTCA)	2023	22	0	VH	5
							40.98–163.9	H	4
							163.9–327.8	M	3
							327.8–573.65	L	2
							573.65–901.46	VL	1
5	AC	Telecommunication (Urban)	Millions	Telecom Regulatory Authority of India (TRAI)	2022	28	4.16–11.04	VH	5
							11.05–59.96	H	4

(continued)

Table 3 (continued)

S.no	C	Indicators	Unit	Data source	Year	W(%)	Class	CR	R
							59.97–76.78	M	3
							76.79–131.82	L	2
							131.83–199.09	VL	1
		Telecommunication (Rural)	Millions				0.95–2.53	VH	5
							2.53–13.74	H	4
							13.74 - 17.59	M	3
							17.59 –30.2	L	2
							30.2 –45.62	VL	1
6	AC	Electricity availability	Megawatt (MW)	National Power Portal (NPP)	2023	32	0 –111	VH	5
							111 –522	H	4
							522 –1056	M	3
							1056 –1856	L	2
							1856 –3942	VL	1
Total						100			
1	E	Natural hazard	Maps	State Disaster Management Authorities (SDMA)	2022	8	Ver High	VH	5
							High	H	4
							Moderate	M	3
2	E	NDVI	Imagery	Satellite data (Landsat 8)	2023	36	-0.62 –0	VH	5
							0 –0.1	H	4
							0.1 –0.2	M	3
							0.2 –0.29	L	2
							0.29 –1	VH	5

(continued)

Table 3 (continued)

S.no	C	Indicators	Unit	Data source	Year	W(%)	Class	CR	R
3	E	Elevation	Imagery	30 m- Shuttle radar topography mission DEM (SRTM) USGS Earth Explorer	2015	22	190 –1242	VL	1
							1242 –2358	L	2
							2358 –3557	M	3
							3557 –4670	H	4
							4670 –6482	VH	5
4	E	Slope	Imagery		2015	19	11.32	VL	1
							11.32 –21.81	L	2
							21.81 –31.53	M	3
							31.53 –42.5	H	4
							42.5 –89.52	VH	5
5	E	Tourism demand	Numbers	SARIMA projections (Indian)	2031	10	-1812	VL	1
							0 –123,231	L	2
							123,232- 201,878	M	3
							201,879 –325,004	H	4
							325,005 –821,510	VH	5
	E	Tourism demand	Numbers	SARIMA projections (Foreign)	2031		-1828	VL	1
							1 –4586.8	L	2
							4588 –7794.2	M	3
							7795 –11,001.6	H	4
							11,003 –14,209	VH	5
6	E	Land use	Imagery	Satellite data (Landsat 8)	2023	5	Agriculture	M	3
							Wasteland	L	2

(continued)

Table 3 (continued)

S.no	C	Indicators	Unit	Data source	Year	W(%)	Class	CR	R
							Built-up	M	3
							Forest	VH	5
							Grass	H	4
							Snow/Glazier	VH	5
							Wetlands/Waterbody	VH	5
Total						100			
1	S	Population density	Person/ sq.km	Census 2011	2011	34	2764	VL	1
							8372	L	2
							1999	M	3
							3679	H	4
							23,396	VH	5
2	S	GSDP	Million	Economic Survey HP 2022–23	2023	12	77,023 –234,268	VL	1
							234,268 –564,165	L	2
							564,165 –652,072	M	3
							652,072 –1,153,890	H	4
							1,153,890 –2,803,653	VH	5
3	S	Marginal workers Male	Population	Census 2011	2011	26	4105 –75,535	VL	1
							75,536 –146,965	L	2
							146,966 –218,395	M	3
							218,396 –289,825	H	4
							289,826 –361,255	VH	5
		Marginal workers Female	Population	Census 2011	2011		0 –1	VL	1

(continued)

Table 3 (continued)

S.no	C	Indicators	Unit	Data source	Year	W(%)	Class	CR	R
							1 –6	L	2
							6 –8	M	3
							8 –12	H	4
							12 –24	VH	5
4	S	Forest area	Hectare	Statistical Abstract HP 2019–20	2020	8	44 –156	VL	1
							156 –397	L	2
							397 –580	M	3
							580 –1682	H	4
							1682 –4566	VH	5
5	S	Net sown area	Hectare	Economic Survey HP 2022–23	2022	8	3397 –8256	VL	1
							8256 –29,815	L	2
							29,815 –41,864	M	3
							41,864 –90,806	H	4
							90,806 –116,800	VH	5
6	S	Water Supply system — Tap	Household	Jal Jeevan Mission –Har Ghar Jal	2023	12	7284 –22,763	VL	1
							22,764 –114,209	L	2
							114,210 –121,658	M	3
							121,659 –171,201	H	4
							171,202 –387,748	VH	5
		Water Supply system –Ground	mgbl min	Central Ground Water Board (CGWB)	2013		5 –9.94	VL	1
							9.94 –23.35	L	2
							23.35 –31.82	M	3

(continued)

Table 3 (continued)

S.no	C	Indicators	Unit	Data source	Year	W(%)	Class	CR	R
							31.82 –40.06	H	4
							40.06 –65	VH	5
			mgbl max				30	VL	1
							30 –115	L	2
							115 –220	M	3
							220 –302	H	4
							302 –432	VH	5
Total						100			

Note Criteria (C), indicators, unit, data source, year of data procurement, weightage percentage, W (%), class, class range (CR), and ranking (R)

Table 4 Criteria-wise consistency ratio (CR)

Criteria	Consistency Index (CI)	Random Index (RI)	Consistency Ratio (CR)
Exposure	0.04	1.24	0.03 (<10%)
Adaptive capacity	0.04	1.24	0.03 (<10%)
Sensitivity	0.02	1.24	0.01 (<10%)

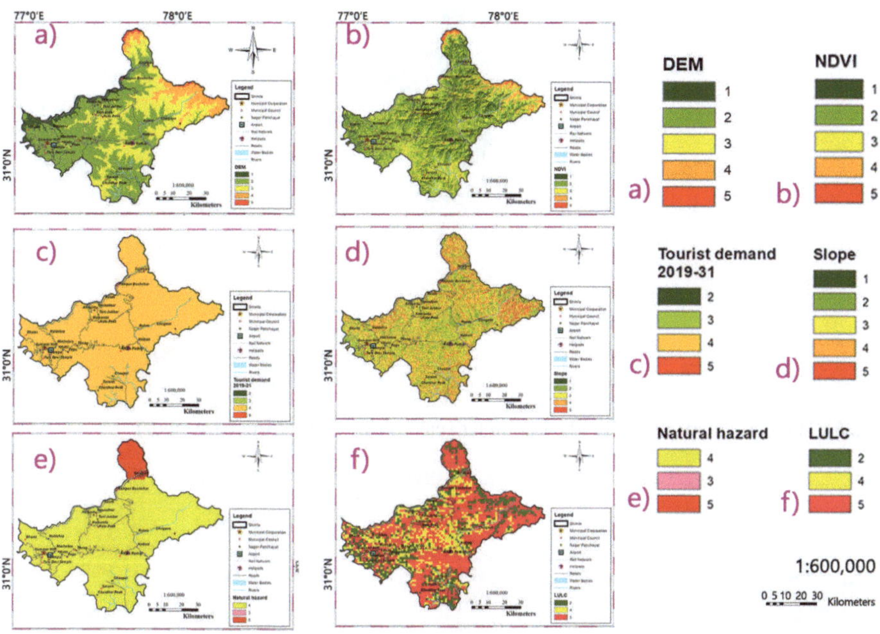

Fig. 4 Thematic maps of each indicator under exposure criteria

The composite tourism vulnerability map of Shimla district is exhibited in Fig. 8, revealing that the northern portions are extremely vulnerable. The upper Shimla has rural characteristics, less population, inadequate infrastructure, dense forest cover, inadequate medical facilities, limited telecommunication services, fewer tourist attractions, limited tourist amenities and high susceptibility to natural hazard. Thus, this region requires restrictive development keeping in mind its susceptibility towards the listed indicators while preparing a strategic plan for boosting tourism. These measures by the policymakers to make tourism sustainable and equitable in the district, allocating funds to develop vulnerable areas, improve quality of life, protect environment and cultural heritage. The significant contribution of the present study is the enlisted indicators, along with their class ranges and weights, to identify areas vulnerable to tourism growth through a systematic methodology.

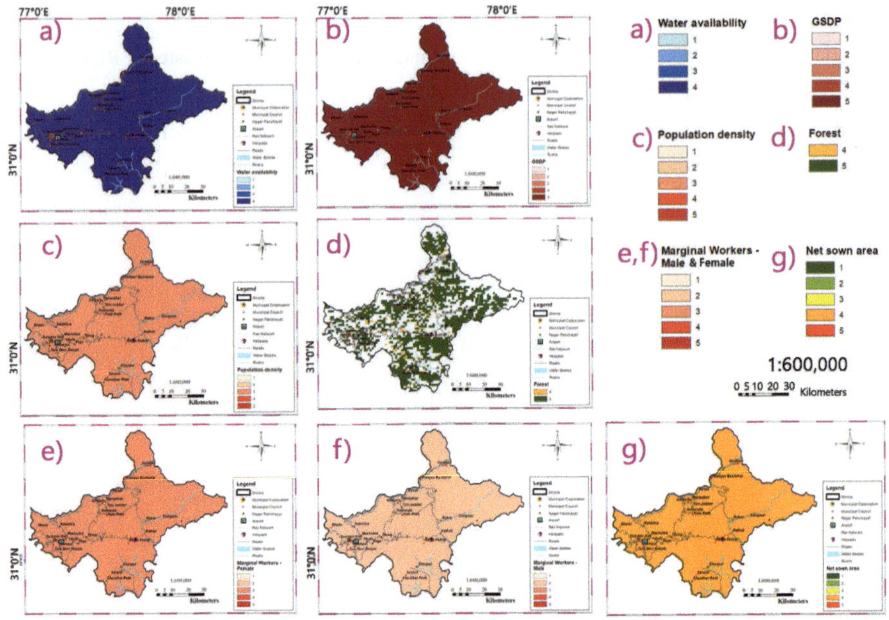

Fig. 5 Thematic maps of each indicator under sensitivity criteria (*Source* Author)

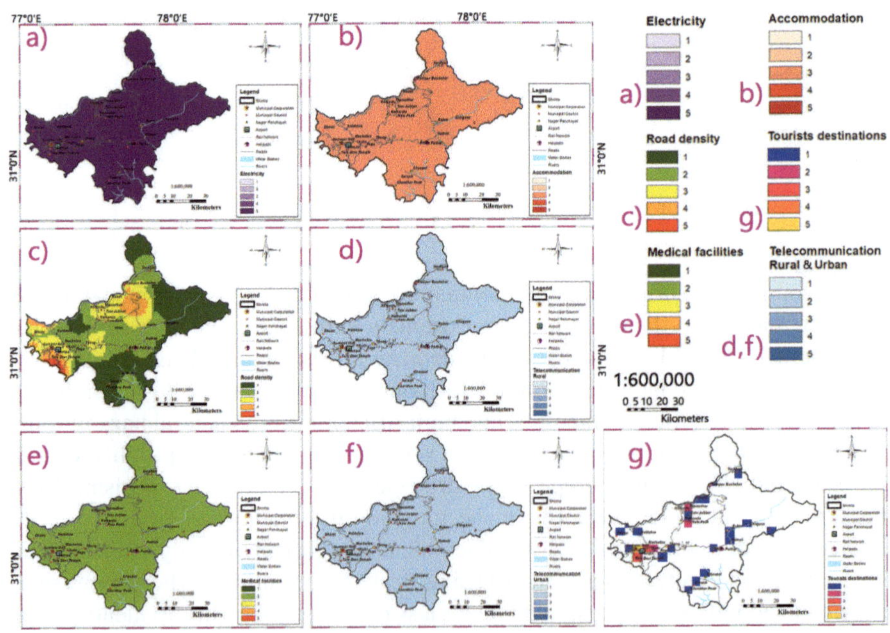

Fig. 6 Thematic maps of each indicator under adaptive capacity criteria (*Source* Author)

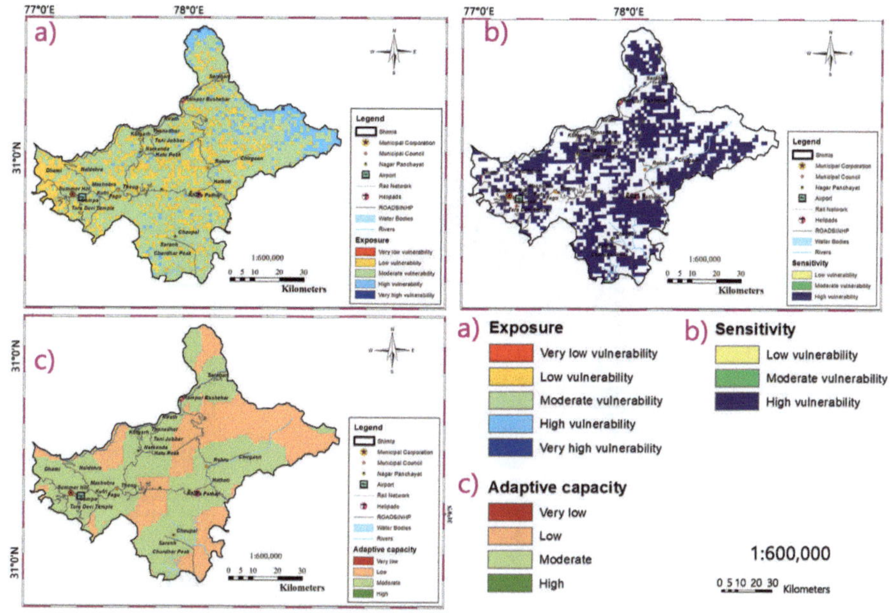

Fig. 7 Cumulative mapping of criteria (*Source* Author)

Table 5 Composite vulnerability statistics of the study area

2 (Low)	3 (Moderate)	4 (High)	Grand total	Area Sq. km	%
202	1501	614	2317	5136	4.17

7 Policy Implications

The policy implications would significantly minimize tourism vulnerability by adopting a context-specific approach for enhancing adaptive capacity. The authorities can intervene through capacity building, stakeholder engagements, and monitoring for accurate and long-term assessment. This study recommended the policy recommendations for reducing tourism vulnerability in the study area by formulating a sustainable tourism development framework, data management for policy adjustments, community empowerment induced with community partnerships and involvement in decision-making, environmental protection, and risk reductions, cultural and heritage preservation, skill development and training for employment, marketing, and promotional strategies to attract high-spending and low impacting tourists, facilitating ease of business and investment in tourist infrastructure.

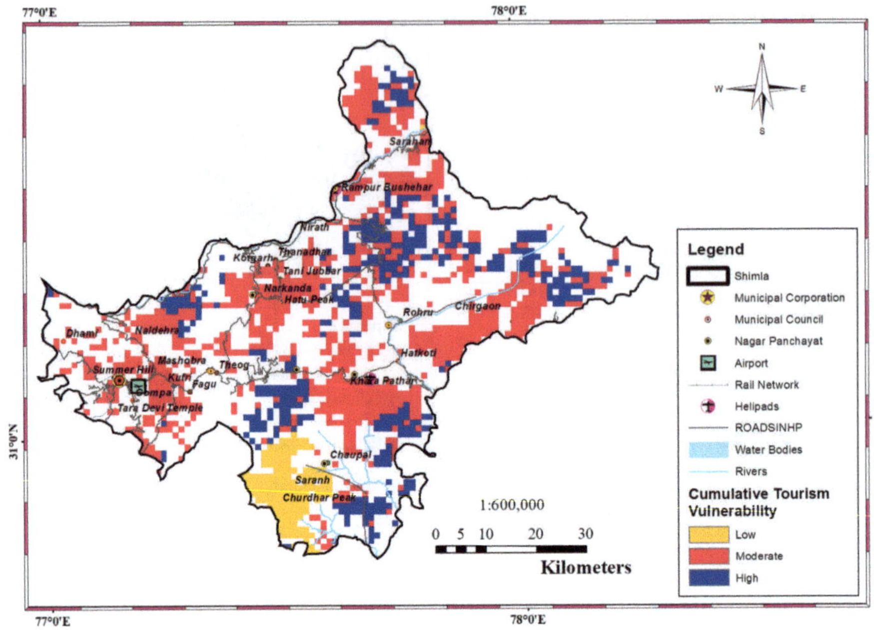

Fig. 8 Composite tourism vulnerability map of Shimla district (*Source* Author)

8 Conclusions

This study focused on identifying and evaluating the indicators influencing tourism vulnerability in Shimla district through spatial mapping and assessment. Scientific assessment of tourism vulnerability has never been conducted by researchers explicitly for Shimla district. Shimla district envisions desaturated destinations through regional tourism diversification, by employing scientific and empirical approach. The novelty of this research is the technique to select appropriate indicators for assessing and mapping tourism vulnerability in hilly region which has multiple susceptibilities. The integrated approach using AHP and expert opinion survey methods were adopted to assist the decision-making. The reliability of the research technique and findings lie on the high consensus between the expert's judgement. The weighted overlay analysis leads to the cumulative mapping of each criterion to spatially represent the level of exposure, degree of sensitivity, and extent of adaptive capacity. Finally, the composite map revealed that Shimla district is highly susceptible toward tourism growth due to its low adaptive capacity, high degree of exposure, and high sensitivity. The results contain implications related to tourism-related activities, which would act as a reference to the stakeholders to ensure investment and fund allocation in the vulnerable areas, improving the quality of life, safeguard local livelihood, protect environment and cultural heritage, and build capacity and awareness to community-based tourism practices.

Though comprehensive, this research has drawbacks such as (1) it relies entirely on the responses collected from the purposive expert sampling approach; (2) the evolution of the indicators was not captured; (3) apart from mapping, there is not enough emphasis on reasoning on the historical events adversely affecting the district; (4) climatic factors such as temperature, rainfall, and relative humidity have not been utilized to assess the degree of exposure; (5) research does not reason behind the moderate and high tourism vulnerability. However, this study opens several future research scopes, such as determining the interrelationship among the indicators of tourism vulnerability, predicting accurate weights of each indicator by employing mathematical models and computer-aided applications, and utilizing high-quality satellite imagery to predict tourism vulnerability. However, future research would depend on data availability to present a micro-level mapping of tourism vulnerability. Also, the index methods such as the Climate change vulnerability Index (CVIT), physical vulnerability index (PVI), and social vulnerability index (SVI) can be adopted for comparative investigation.

References

1. UNWTO (2023) World Tourism Barometer. World Tour Barom, 21(1), pp. 1–6
2. Manisha K, Singh I (2024) Forecasting of Indian and foreign tourist arrivals to Himachal Pradesh using Decomposition, Box-Jenkins, and Holt-Winters exponential smoothing methods. Asia-Pac J RegNal Sci. https://doi.org/10.1007/s41685-024-00344-8
3. Ng SI, Chia KW, Ho JA, Ramachandran S (2017) Seeking tourism sustainability—A case study of Tioman Island, Malaysia. Tour Manag 58:101–107. https://doi.org/10.1016/j.tourman.2016.10.007
4. Oad S, Jinliang Q, Shah SBH, S. ul R (2022) Memon, Tourism: economic development without increasing CO2 emissions in Pakistan. Environ Dev Sustain, 24(3), pp. 4000–4023, 2022, https://doi.org/10.1007/s10668-021-01601-y
5. Sahoo B, Bhaskaran PK (2018) Multi-hazard risk assessment of coastal vulnerability from tropical cyclones—A GIS based approach for the Odisha coast. J Environ Manage 206:1166–1178. https://doi.org/10.1016/j.jenvman.2017.10.075
6. Bellert F et al. (2021) Application of Remote Sensing and GIS for Risk Assessment in Monastir, Tunisia BT—Environmental remote sensing and GIS in Tunisia. Khebour Allouche F, Negm AM Eds. Cham: Springer International Publishing, pp 191–210. https://doi.org/10.1007/978-3-030-63668-5_10
7. Bagheri M et al (2021) Application of multi-criteria decision-making model and expert choice software for coastal city vulnerability evaluation. Urban Science 5(4):84. https://doi.org/10.3390/urbansci5040084
8. Manisha K, Singh I, Chettry V (2023) Investigating and analyzing the causality amid tourism, environment, economy, energy consumption, and carbon emissions using Toda–Yamamoto approach for Himachal Pradesh, India. Environ Dev Sustain, no. 0123456789. https://doi.org/10.1007/s10668-023-04252-3
9. D. of S. & T. Government of India, National mission for sustaining the himalayan ecosystem under national action plan on climate change mission. (2010). Available: https://dst.gov.in/sites/default/files/NMSHE_Mission_document.pdf
10. Deng Y (2022) Spatial Analysis on the vulnerability of Tourism economic system based on bp neural network: the guangdong-Hong Kong-Macao Greater Bay Area (GBA). Math Probl Eng, 2022, https://doi.org/10.1155/2022/5506388

11. Perch-Nielsen SL (2010) The vulnerability of beach tourism to climate change-an index approach. Clim Change 100(3):579–606. https://doi.org/10.1007/s10584-009-9692-1
12. AKC, Thapa Parajuli RB (2014) Tourism and its impact on livelihood in Manaslu conservation area, Nepal. Environ Dev Sustain, 16(5), pp 1053–1063, https://doi.org/10.1007/s10668-013-9512-7
13. Qiu H, Fan DXF, Lyu J, Lin PMC, Jenkins CL (2019) Analyzing the economic sustainability of tourism development: evidence from Hong Kong. J Hosp Tour Res 43(2):226–248. https://doi.org/10.1177/1096348018777046
14. Manisha K, Singh I (2024) Investigating green economy studies using a bibliometric analysis, no. 0123456789. Springer US, https://doi.org/10.1007/s13132-024-02237-9
15. Navale S, Bhagat V (2022) Use of AHP based weighted analysis for impact assessment of coastal tourism in ratnagiri district, Maharashtra (India): respondents' point of view. J Geogr Stud 6(1):1–32. https://doi.org/10.21523/gcj5.22060101
16. Santos-Lacueva, R, Clavé SA, Saladié Ò (2017) The vulnerability of coastal tourism destinations to climate change: The usefulness of policy analysis. Sustainability (Switzerland), 9(11), https://doi.org/10.3390/su9112062
17. Shi H, Shi T, Liu Q, Wang Z (2021) Ecological vulnerability of tourism scenic spots: Based on remote sensing ecological index. Pol J Environ Stud 30(4):3231–3248. https://doi.org/10.15244/pjoes/129916
18. Barbhuiya MR, Chatterjee D (2020) Vulnerability and resilience of the tourism sector in India: Effects of natural disasters and internal conflict. Tour Manag Perspect, 33(2019), p 100616, https://doi.org/10.1016/j.tmp.2019.100616
19. Barbhuiya MR (2021) Overtourism in Indian cities: a case study of Nainital. Int J Tour Cities 7(3):702–724. https://doi.org/10.1108/IJTC-08-2019-0148
20. Espiner S, Becken S (2014) Tourist towns on the edge: Conceptualising vulnerability and resilience in a protected area tourism system. J Sustain Tour 22(4):646–665. https://doi.org/10.1080/09669582.2013.855222
21. Sahani N (2020) Application of analytical hierarchy process and GIS for ecotourism potentiality mapping in Kullu District, Himachal Pradesh, India. Environ Dev Sustain 22(7):6187–6211. https://doi.org/10.1007/s10668-019-00470-w
22. Anfuso G, Postacchini M, Di Luccio D, Benassai G (2021) Coastal sensitivity/vulnerability characterization and adaptation strategies: A review. J Mar Sci Eng 9(1):1–29. https://doi.org/10.3390/jmse9010072
23. Saaty TL (2001) Fundamentals of the analytic hierarchy process BT—The analytic hierarchy process in natural resource and environmental decision making. Schmoldt DL, Kangas J, Mendoza GA, Pesonen M Eds, Dordrecht: Springer Netherlands, pp 15–35. https://doi.org/10.1007/978-94-015-9799-9_2

Safeguarding the Western Himalayas: Sustainable Disaster Management in the Hilly Regions of Himachal Pradesh

Shubham Sharma, Shalini Sharma, and Swasti Sharma

1 Introduction

Disasters, as defined extreme sudden catastrophic events that occur in a specific area, causing widespread damage to both human prosperity and ecosystems. These events can be broadly categorized into two types based on their causes: natural disasters and anthropogenic disasters. Natural disasters are those triggered by natural forces, such as earthquakes, floods, or wildfires, while anthropogenic disasters result from human activities, like industrial accidents or environmental pollution. [1] India has suffered an economic loss of $4.2 billion in 2022 due to extreme weather events and climatic disasters, most of which were caused by flood, according to a new World Meteorological Organization (WMO) report [2].

Disasters and Disaster Management in Hilly Areas

Hilly regions, such as the western Himalayas, are particularly vulnerable to natural disasters due to their unique topographical and climatic conditions. Steep slopes, fragile geological formations, and intense rainfall contribute to the frequent occurrence of landslides, flash floods, and avalanches. The isolation and inaccessibility of many hilly regions further complicate disaster response and relief efforts. Effective disaster management in hilly areas requires a multifaceted approach that includes risk assessment, early warning systems, infrastructure development, and community engagement. Risk assessment involves identifying and mapping hazard-prone

S. Sharma · S. Sharma
Indian Institute of Technology, Kharagpur, India
e-mail: shubham_s@ar.iitr.ac.in

S. Sharma
e-mail: shalini.sharma173@gmail.com

S. Sharma (✉)
Amity School of Architecture and Planning, Amity University, Punjab, India
e-mail: 2020rar9574@mnit.ac.in

V. Chettry et al. (eds.), *Geoinformatics for Hilly Regions*, Advances in 21st Century Human Settlements, https://doi.org/10.1007/978-981-95-0156-4_12

areas, understanding the vulnerabilities of communities, and evaluating the potential impacts of disasters. Infrastructure development should focus on constructing resilient buildings, roads, and bridges that can withstand natural hazards. This includes using appropriate construction materials and techniques, as well as incorporating disaster-resistant features into design and planning. In hilly regions, it is also crucial to maintain and enhance natural vegetation and forests, which play a vital role in stabilizing slopes and reducing the risk of landslides and erosion. Community engagement is also a critical component of disaster management in hilly areas. Local communities often have valuable traditional knowledge and practices that can contribute to disaster risk reduction. Involving community members in planning and decision-making processes helps ensure that disaster management strategies are culturally appropriate and locally relevant. Additionally, community-based disaster preparedness programs can enhance the resilience of communities by raising awareness, building local capacity, and fostering a culture of safety and preparedness.

This study explores the complex interplay between disaster resilience and sustainable development in Himachal Pradesh. By employing advanced methodologies like Shannon's Entropy and Fuzzy TOPSIS, the research aims to deepen understanding of flood vulnerability and advocate for integrating disaster risk reduction principles into mainstream development initiatives. Emphasizing a holistic disaster management continuum and a community-based approach, the study provides a comprehensive framework for enhancing resilience and promoting sustainable development across the region.

1.1 General

Himachal Pradesh, situated in the western Himalayas, attained statehood in 1971. Bordered by the states of Haryana, Uttarakhand, Punjab, and Uttar Pradesh, it shares its boundaries with the Union Territory of Ladakh. It also shares international borders with Tibet in tribal areas of Kinnaur and Spiti. This scenic region is famed for its captivating scenery and varied ecosystems, but it remains vulnerable to natural calamities, including flash floods, landslides, and avalanches, which underscores the importance of proactive disaster management strategies in the region. As the state continues to evolve, the need for effective disaster preparedness and response becomes increasingly crucial to safeguard both human lives and the fragile mountainous environment.

1.2 Administration

Himachal Pradesh spans an area of 55,673 square kilometers and is home to 6,864,602 inhabitants. The state is divided into 12 districts, which fall under three administrative

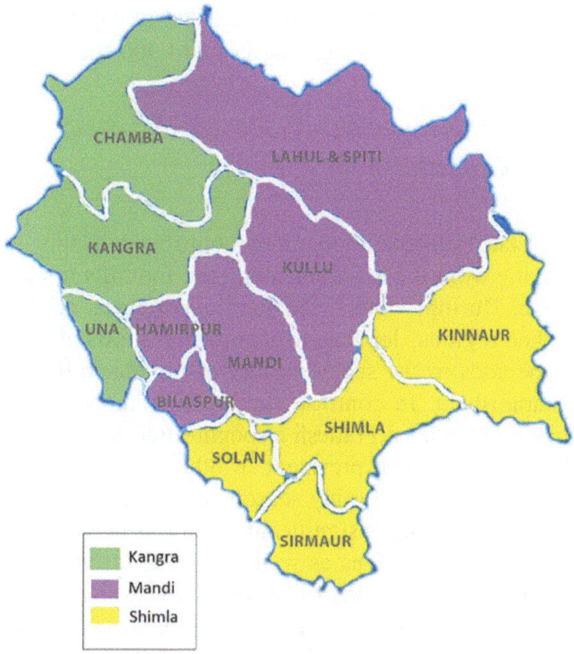

Fig. 1 Administrative divisions of Himachal Pradesh (*Source* Author)

divisions: Kangra, Mandi, and Shimla (as illustrated in Fig. 1). The districts under each division are as follows: –

- **Kangra**: Kangra, Chamba, Una
- **Mandi:** Hamirpur, Bilaspur, Kullu, Mandi, Lahaul and Spiti
- **Shimla:** Sirmaur, Shimla, Solan, Kinnaur

Himachal Pradesh has 172 tehsils, 54 urban local bodies (ULBs), and 3,226 g panchayats (GPs), forming a comprehensive administrative network that facilitates inclusive governance and ensures the effective delivery of services to its diverse [3].

1.3 Geography

Himachal Pradesh lies between 450 and 7000 m above mean sea level. The elevation increases from West to East and from South to North. Its landscape is a complex blend of mountains, hills, valleys, and glens. Notable mountain ranges include the Siwaliks, Dhauladhar, Pir Panjal, Great Himalayan, and Zanskar. The state is part of the Himalayan Mountain range, and it is usually split into three geographical areas: the 'Siwaliks', the 'Lesser and Middle Himalayas' and the 'Greater Himalayas'. The lower Himalayan region experiences mean temperatures ranging between 16–20 degrees Celsius, whereas the upper Himalayan areas record temperatures below

8 degrees Celsius. Approximately 25% of the state's land area is covered by forests, amounting to a total forest cover of 14,300 square kilometers.

1.4 Climate

The state's diverse topography contributes to this climatic variety, with elevations ranging from the lowlands along the foothills to imposing peaks in the Himalayan range. During winters, the state experiences cold temperatures, especially in the higher regions, leading to significant snowfall in many areas.This snowy landscape adds to the scenic beauty of popular hill stations like Shimla, Manali, and Dharamshala. In contrast, the summer months are generally mild and pleasant, making Himachal Pradesh a sought-after destination for those seeking respite from the scorching heat prevalent in other parts of India. The average annual rainfall is 1,251(mm). The rainy season start at the end of the month of June. The state's topography exhibits significant variations in elevation, leading to marked disparities in climatic conditions as well and till today it faces numerous flood events. Flash floods, cloudbursts, landslides and snow avalanche are the common natural disasters prevalent the state. The total loss caused by disaster in state in 2023 was 12,000 cr as per data collected by DDMA. This study focuses on understanding the various disasters prevalent in Himachal Pradesh and the scope for disaster management strategies within the framework of sustainable development in the state.

2 Research Design

The research design for this study followed a systematic approach, taking Kullu district as a case study to comprehensively understand the disaster risk and vulnerability landscape of Himachal Pradesh. The initial phase Involved an extensive literature review through research papers, government reports and newspaper data, delving into existing studies on disaster risk, vulnerability, and management, with a keen emphasis on methodologies applied to similar regions. Geospatial analysis played a pivotal role in the research design, leveraging data such as Digital Elevation Models (DEM) (see Fig. 2) and topographic maps to gain insights into the physical geography and landscape features and overall vulnerability map of Himachal Pradesh (see Fig. 2). Geographical information system (GIS) tools were employed to map and visualize vulnerable areas, contributing to a spatially informed understanding.

Shannon's entropy method was applied to assess information entropy within physical attributes associated with flood vulnerability. This involved quantifying uncertainty and disorder related to land topography, hydrological conditions, and infrastructure, offering a nuanced understanding of vulnerability factors. Furthermore, the Fuzzy Technique for Order of Preference by Similarity to Ideal Solution (Fuzzy TOPSIS) was employed for a comprehensive evaluation of physical vulnerability by

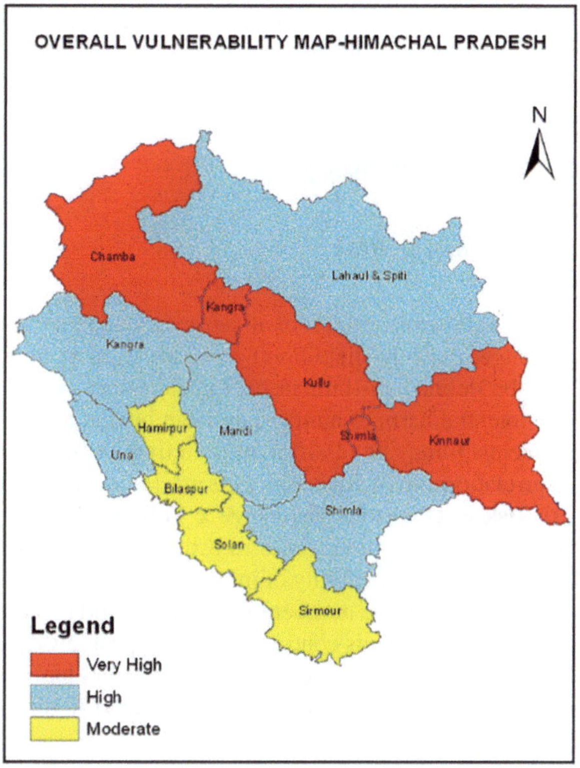

Fig. 2 Overall vulnerability map of Himachal Pradesh (*Source* [8])

considering multiple criteria, including elevation, land use, and drainage systems. The integration of diverse criteria provided a holistic ranking of areas based on their susceptibility to flooding. In addition to these quantitative methods, the study incorporated qualitative insights through interviews with a panel of experts from various government departments. These experts contributed valuable perspectives on parameters associated with physical vulnerability, such as drainage systems, proximity to rivers, slope, and elevation. Additionally, an examination of the state policy on disaster management was conducted to further understand the regulatory framework in place.

The culmination of these methodologies enabled the creation of a vulnerability map through weighted overlay analysis. The map serves as a visual representation of high-risk areas, providing decision-makers with insights to guide targeted interventions and resource allocation. A comprehensive map encompassing the entire state for the analysis of vulnerable areas to various disasters was not feasible within the constraints of time. Consequently, a focused approach was adopted, selecting a single district and a specific type of disaster for in-depth study. While the study's focal point was the Kullu district, the methodology demonstrates scalability and potential applicability to broader geographical contexts within Himachal Pradesh, and other hilly regions with the same climatic and topographical complexities, offering a foundation for comprehensive disaster risk assessment and management strategies.

3 Literature Study

3.1 Disaster Risk

The total Potential effects of all hazards, frequency, the number of persons and assets at danger, and the likelihood of destruction is known as Disaster Risk. Hazards may be defined as 'human ecological interaction that can generate disaster. It is the potential harm to humans, and can be due to natural, physical or environmental elements. It may be of different time frequencies as well—chronic or episodic. The interrelationship of Risk with Hazard, Exposure and Vulnerability can be expressed as: **Risk = Hazard x Exposure x Vulnerability** [5]. As much as the intensity of the hazard event plays a role in estimating the losses and effects of a disaster, the exposure and vulnerability of individuals and locations sometimes play a bigger role. [6] suggests that the number of disasters could rise from about 400 in 2015 to 560 in 2030 if current trends continue. Over the past few decades, the average annual direct economic damage caused by natural disasters has increased by around 145%, from about $70 billion in the 1990s to just over $170 billion in the 2010s.

3.2 Vulnerability

The term "vulnerability" assumes multiple definitions, reflecting its comprehensive nature. Broadly, vulnerability pertains to the extent to which a community, populace, or system is predisposed to the adverse consequences of a disaster event.

Vulnerability is an intricate concept encompassing diverse facets, including social, economic, environmental, and physical elements that collectively heighten the susceptibility of a population or system to the detrimental impacts of disasters. Gaining a profound understanding of vulnerability stands as a pivotal component of effective disaster management. Such comprehension serves to pinpoint the specific populations and systems most vulnerable, thus facilitating prioritized interventions aimed at diminishing vulnerability and bolstering resilience.

Within the domain of disaster management, [7]discerns three principal schools of thought:

- **The Hazard Approach**: This theoretical framework directs its attention to the physical attributes intrinsic to a disaster event, encompassing aspects such as its intensity, frequency, and duration. The hazard approach underscores the imperative of anticipatory measures and the amelioration of the consequences of natural disasters through scientific and technical methodologies. This involves enhancements in building codes, the development of early warning systems, and the strategic implementation of land-use planning measures.

- **The Vulnerability Approach**: This paradigm places its primary emphasis on the socio-economic determinants contributing to the vulnerability of both individuals and communities when confronted with disasters. Recognizing the inherent disparities in susceptibility among various segments of society, the vulnerability approach underscores the necessity of addressing fundamental issues such as poverty, inequality, and social exclusion. These efforts are crucial in reducing vulnerability levels and concurrently fostering resilience.
- **The Capacity Approach:** This approach focuses on the ability of individuals and 15 communities to prepare for, respond to, and recover from disasters. The capacity approach emphasizes the importance of empowering individuals and communities to take an active role in disaster management and recognizes that effective disaster management requires the involvement of a range of stakeholders, including government agencies, civil society organizations, and the private sector.

3.3 Geographical Features of the Area

Himachal Pradesh is classified into zones based on elevation above mean sea level, encompassing both high and low regions. The western and southern areas, including parts of Una, Bilaspur, Hamirpur, Kangra, Solan, and Sirmaur districts, lie below 1000 meters and experience a tropical to subtropical climate with corresponding vegetation. In contrast, some portions of Solan, Sirmaur, Mandi, Chamba, and Shimla districts are situated at elevations between 1000 and 3500 meters, where the climate varies from subtropical to wet-temperate with increasing altitude. The regions of Lahaul Spiti, Kullu, and Kinnaur, ranging between 3500 and 6700 meters, fall into the dry-temperate, subalpine, and alpine zones, characterized by very sparse rainfall. The digital elevation model (DEM), along with district boundaries, provides a clearer understanding of the state's diverse topography (as illustrated in Fig. 3).

4 Study of Different Disasters Across HP

The GOI identifies 33 types of hazards, and Himachal Pradesh (HP) is susceptible to 25. Along with this, HP is also facing the threat posed by climate change. The man–animal conflict and man-made disruptions only add to the vulnerability of the state. The overall vulnerability map of Himachal has been shown in Fig. 2. The following events occur due to hydro-meteorological disasters and geologically related disasters.

Hydro-meteorological disasters.

1. Cloud Burst
2. Floods

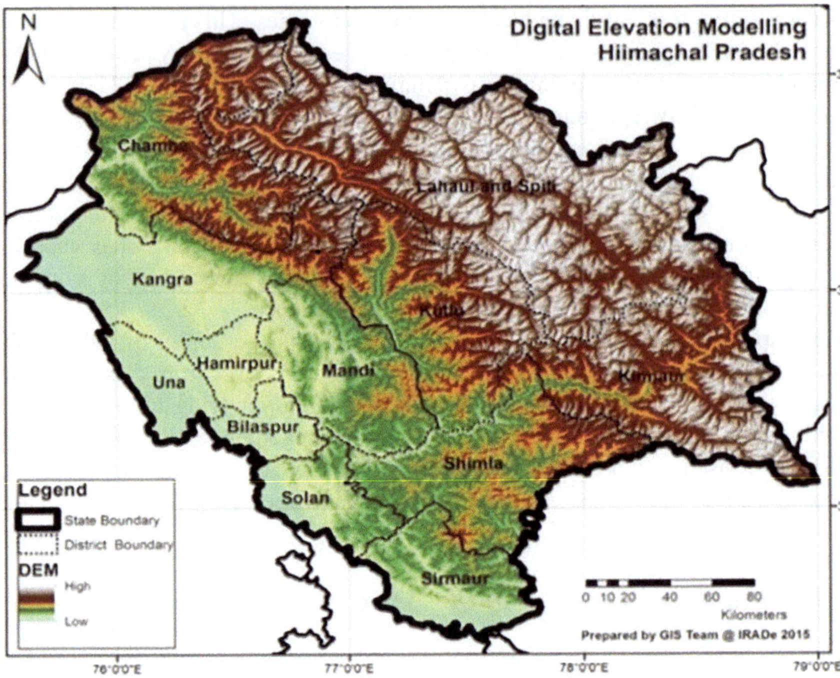

Fig. 3 Digital elevation modelling of Himachal Pradesh (*Source* [14])

3. Heat Wave and Cold Wave
4. Droughts
5. Hailstorm
6. Snow Avalanches
7. Thunder and Lightning

Geologically Related Disasters.

1. Landslides and Mudflows
2. Earthquakes

The severity of these hazards in various districts within the state have been outlined in Table 1.

5 Methodology

The study has been conducted in two parts. The first part of the study consists of the literature review on disaster risk, disaster management in hilly areas, vulnerability, etc. and the second part of the study includes interview survey and entropy

Table 1 District-wise hazard threat in Himachal Pradesh (Source: [8])

District	Floods	Cloud Burst	Avalanche	Drought	Eatlhquake	Latldslide
Kangra	M	M	M	H	VH	L
Chamba	H	H	M	M	VH	VH
Hanlirpur	L	L		M	H	L
Mandi	H	H		M	VH	H
Kullu	H	VH	H	M	VH	VH
Bilaspur	L	L		M	H	M
Una	H	L		H	H	L
Simour	L	M		M	H	L
Solan	L	L		M	H	M
Kilmaur	H	VH	VH	M	H	H
Lahaul and Spiti	M	H	VH	M	H	M
Shimla	H	H	M	M	VH	H

(Symbol: L= Low. M= Meditun. H= High. Vll= Very High)

weight method (EWM) analysis. In the conducted study, a panel of eight individuals participated in personal interviews. These members were experts from various government departments like Town & country planning, Department of Environment, Science & Technology, District Disaster Management Authority. The questions revolved around examining how parameters associated with physical vulnerability, such as the existence of drainage systems, proximity to rivers, slope, and elevation, influence the vulnerability to flooding. The participants were tasked with rating predefined criteria on a scale ranging from 1 to 5, leveraging their expertise and knowledge in the field. To systematically analyze and synthesize the gathered information, method of Entropy has been used. The entropy weight method (EWM) is a significant information weighting model that has been widely studied and applied. Unlike various subjective weighting models, EWM's key advantage lies in its ability to minimize human interference in determining indicator weights, thereby enhancing the objectivity of the overall evaluation results [9].

6 Results and Discussions

The method used of entropy have been listed below, and final result is shown in Table 2.

After obtaining the weights for the criteria, Fuzzy TOPSIS method was applied for further analysis. Fuzzy TOPSIS provides a systematic and objective approach to evaluating alternatives based on multiple criteria. In this method, the optimal alternative is identified as the one closest to the fuzzy positive ideal solution and farthest from the fuzzy negative ideal solution. [10] The first step in Fuzzy TOPSIS method

Table 2 Result of Entropy method

	Presence of drain lines	Proximity to river	Slope (hilly terrain)	Spot elevation	Proximity to major traffic routes	Topographic wetness index
weight	0.385	0.077	0.023	0.108	0.287	0.12
	Positive	Negative	Positive	Positive	Positive	Positive
	Presence = high score	Shorter distance/ more canals = lower score	Presence = high score	Presence = high score	Presence = high score	Presence = high score

involved constructing a fuzzy matrix using the assigned weights to comprehensively represent the decision-making criteria. Following this, a normalization process was applied to the fuzzy matrix, ensuring that the values were scaled appropriately to facilitate meaningful comparisons. The weights derived were normalized concerning the fuzzy matrix, aligning them with the overall decision framework. Following this, the distances from the ideal positive and negative solutions were calculated, [11]offering insights into how well each alternative performed relative to the optimal and least favorable outcomes. This distance calculation provided a quantitative measure of the alternatives' proximity to the ideal solutions. Finally, a comprehensive ranking of the alternatives was executed based on the calculated distances, allowing for a systematic evaluation and prioritization of the available options by the established criteria and weights. This multi-step process served as a structured approach to decision-making, integrating fuzzy logic principles and normalization techniques to derive a robust ranking of alternatives, as outlined in Table 3. The closeness coefficient CC1 represents the distances to fuzzy positive ideal solution. This coefficient of each alternative is calculated as:

$$\mathrm{CC1} = \mathrm{di} - /\mathrm{di} - +\mathrm{di}* \quad (1)$$

Result of responded D were taken into consideration. Then using weighted overlay analysis, the mapping of existing physical conditions was done & final vulnerability

Table 3 Ranking obtained after TOPIS

	cc-	rank
A	0.289766407	3
B	0.306174753	2
C	0.171089595	7
0	0.344453108	1
E	0.240954779	5
F	0.122278312	8
G	0.192909622	6
H	0.258445978	4

map was prepared (Fig. 4). Weighted overlay analysis can be used in disaster management to identify areas that are at high risk for different types of disasters and to prioritize interventions to reduce those risks. The weighted overlay analysis, a critical component of the methodology, integrates diverse spatial data, including slope, rainfall, elevation, distance from the river, and topographic wetness index (TWI). This approach allows for a holistic assessment of physical vulnerability, considering the intricate interplay of environmental factors.

The vulnerability map generated through spatial analysis serves as a powerful tool for decision-makers, offering a visual representation of high-risk areas that can guide targeted interventions and resource allocation.

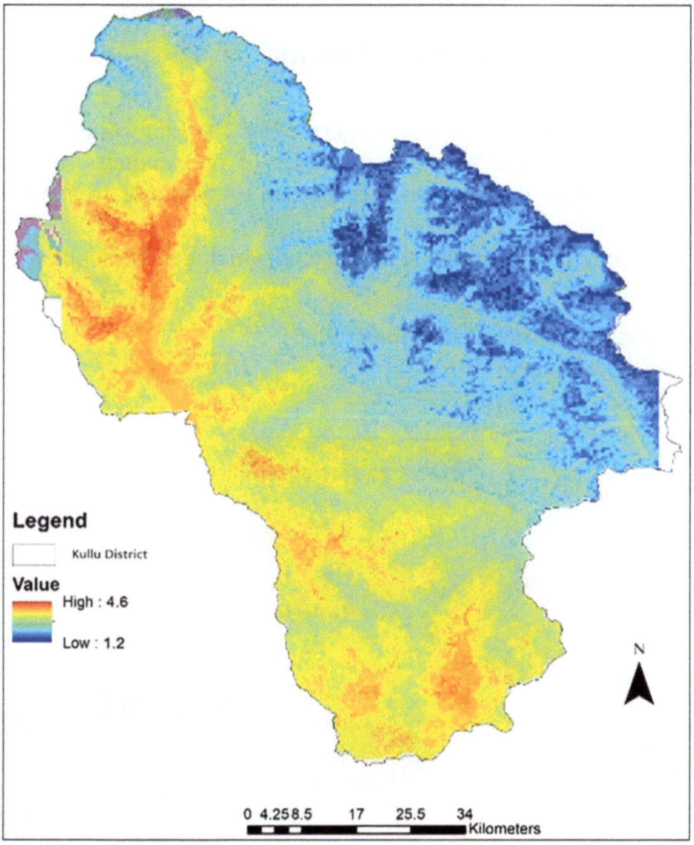

Fig. 4 Vulnerability map of Kullu district (*Source*: Author)

6.1 State Policy on Disaster Management

The Himachal Pradesh State Policy on Disaster Management was published in 2012. The State Policy, which is based primarily on the National Policy on Disaster Management 2009, establishes a comprehensive plan for disaster risk management for the State (NPDM). According to policy, the Himachal Pradesh disaster management vision is to.

> Build a safe and disaster resilient Himachal Pradesh by developing a holistic, proactive, technology-driven, and community-based strategy through a culture of prevention, mitigation, preparedness and response.

6.2 Disaster Management Continuum

The State Policy identifies six components (see Fig. 5) of the continuum of disaster management, including the pre-disaster phases of preparedness, mitigation, and prevention, as well as the post-disaster phases of response, rehabilitation, and recovery.

The State Policy has identified some actions which can be adopted for disaster readiness, including.

(a) Preparation of Disaster Management Plans
(b) Capacity Development

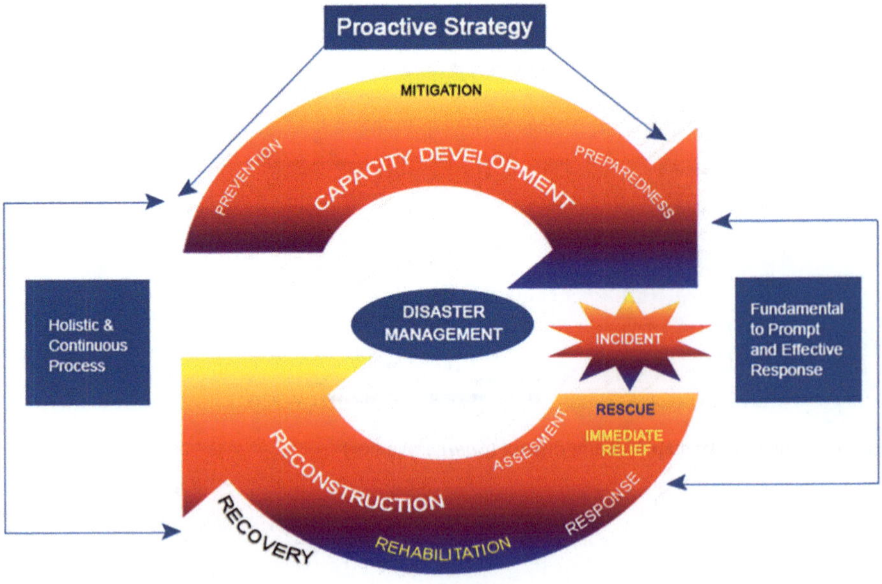

Fig. 5 Disaster management continuum (*Source* [8])

(c) Forecasting and Early Warning Systems
(d) Communications and Information Technology Support
(e) Setting up and Strengthening of Emergency Operations Centre
(f) Training, Simulation, and Mock Drills
(g) Medical Preparedness and Mass Casualty Management
(h) Community-Based Disaster Preparedness

6.3 Mainstreaming Disaster Risk Reduction for Sustainable Development in Himachal Pradesh

The integration of Disaster Risk Reduction (DRR) principles into development programs, activities, and projects has emerged as a powerful strategy for Himachal Pradesh to strengthen its resilience to disasters. This approach eliminates the need for separate DRR funding, as existing allocations are designed such that risk variables are taken into account to safeguard development ventures from the disasters risks and prevent the creation of new risks of disasters as a result of development, either knowingly or unknowingly. All programs, events, and initiatives must be reviewed by each department to determine which ones need to be mainstreamed. Under the direction of subject matter specialists in the relevant fields, the mainstreaming exercise must be carried out methodically.

Steps to be taken for sustainable development with DRR.

- Stabilizing crucial slopes in Shimla and other significant towns/villages that have been chosen.
- Preventing disasters from endangering monuments and historic structures.
- Reducing the risk of landslides on a few state highways.
- Creating a forest fire early warning system.
- Building large hospitals' capacity for managing mass casualties.
- Retrofitting earthquake-prone buildings to reduce their danger

7 Conclusion

The research delves into the comprehensive analysis of flood vulnerability in the Kullu district of Himachal Pradesh, employing a methodology integrating statistical techniques, expert opinions, and spatial analysis tools. The study showcases the robustness of the applied methods, including Shannon's Entropy for weight determination, Fuzzy TOPSIS for comprehensive ranking, and weighted overlay analysis for mapping physical conditions. The vulnerability map generated serves as a valuable tool for decision-makers, aiding in the identification of high-risk areas and the allocation of resources for targeted interventions.

The paper not only highlights the specific vulnerabilities in the Kullu district but also suggests the potential scalability of the methodology to encompass the entire state of Himachal Pradesh. Despite the practical constraints limiting the study's scope, the groundwork laid in the smaller area of Kullu district provides a foundation for future research endeavors on a larger scale. The adaptability of the methodology to various other disasters, such as landslides, earthquakes, and droughts, emphasizes its broader applicability for informing comprehensive disaster risk management strategies.

Furthermore, the study explores the current disaster management scenario in Himachal Pradesh, emphasizing the importance of mainstreaming disaster risk reduction (DRR) principles into development programs for sustainable and resilient development. The State Policy on Disaster Management, along with its identified components across the disaster management continuum, serves as a guide for building a disaster-resilient Himachal Pradesh.

In essence, the research contributes not only to the understanding of flood vulnerability but also to the broader discourse on sustainable disaster management. By incorporating both localized insights and the potential for wider application, the study provides a holistic perspective for informed decision-making in disaster risk reduction and management in the region and beyond.

The study's methodology has several limitations and opportunities for future research. One key limitation is the focus on a relatively small geographical area, the Kullu district, which, although practical, may not capture the full diversity of flood vulnerability factors present across the entire state of Himachal Pradesh. Additionally, the reliance on expert opinions for weight determination introduces subjective bias, which could be mitigated through a more extensive data-driven approach. Future research could expand the scope of the study to include the entire state, enhancing the scalability and applicability of the methodology. This expansion would allow for a more comprehensive disaster risk management strategy. Moreover, the methodology's adaptability suggests its potential application to other disasters such as landslides, earthquakes, and droughts, thereby broadening its impact and utility in disaster risk assessment and management across diverse geographical contexts.

References

1. Chandra U (1992) Seismotectonics of Himalayas. Current Science, 62, pp 40–71. EM-DAT, 2006, The OFDA/CRED International Disaster Database (Brussels: Universite´ Catholique de Louvain)
2. Chanchitpricha C, Bond A (2018) Developing a sustainability assessment framework for thailand's climate change master plan. In: 38th annual conference of the international association for impact assessment 16–19 May 2018 | Durban Intl. Convention Center | Durban | South Africa, May, 1–5. https://conferences.iaia.org/2018/final-papers/Chanchitpricha, Chaunjit - Developing a Sustainability Assessment framework for Thailand's Climate Change Master Plan.pdf
3. Creating contours using ArcMap/ArcScene | Map and Data Library. (n.d.). Retrieved September 26, 2021, from https://mdl.library.utoronto.ca/technology/tutorials/creatingcontours-using-arcmaparcscene 9)
4. https://commons.wikimedia.org/wiki/File:Divisions_of_Himachal_Pradesh.png

5. Cutter SL (1996) Vulnerability to environmental hazards. Prog Hum Geogr 20(4):529–539. https://doi.org/10.1177/030913259602000407
6. Deepak S, Rajan G, Jairaj PG (2020) Geospatial approach for assessment of vulnerability to flood in local self governments. Geoenvironmental Disasters, 7(1). https://doi.org/10.1186/s40677-020-00172-w 11)
7. Demir V, Kisi O (2016) Flood hazard mapping by using geographic information system and hydraulic model: Mert River, Samsun, Turkey. Adv Meteorol, (2016). https://doi.org/10.1155/2016/4891015
8. HP State Council for Environment Science and Technology https://himcoste.hp.gov.in/
9. Global Assessment Report on Disaster Risk Reduction (GAR) (2022)
10. Department of Environment, Science and Technology (2012) .State strategy and action plan on climate change, Himachal Pradesh
11. Thakur VC (2004) Active tectonics of Himalayan frontal thrust and seismic hazard to Ganga Plain. Curr Sci 86(11):1554–1560
12. Disaster management plan - hpsdma.nic.in. Available at: https://www.hpsdma.nic.in/WriteReadData/LINKS/6e41df0d-ae74-4954-8d3b-7d4e145a7365.pdf (Accessed: October 10, 2022).
13. Pelling M (2003) Natural disaster and development in a Globalizing World. Available at: https://doi.org/10.4324/9780203402375
14. Socio economic vulnerability of Himachal Pradesh to climate https://irade.org/Socio Economic Vulnerability of Himachal Pradesh to Climate Change_30.03.2015.pdf
15. Disaster Management Act 2005
16. Govt of H.P. Revenue Department
17. Government of Netherlands, (2021). National Delta Programme 2022. https://english.deltaprogramma.nl/
18. Jha Madan Kumar. Natural and anthropogenic disaster: An overview, Springer, 2010, 1–16
19. Hanna A (2019) Rio de Janeiro's Landslide Mitigation Tactics, Part 1: A Technical History and Overview - RioOnWatch. Rio on Watch Website. https://rioonwatch.org/?p=54009
20. Hoang LP, Biesbroek R, Tri VPD, Kummu M, van Vliet MTH, Leemans R, Kabat P, Ludwig F (2018) Managing flood risks in the Mekong Delta: How to address emerging challenges under climate change and socioeconomic developments. Ambio 47(6):635–649. https://doi.org/10.1007/s13280-017-1009-420)
21. Hooper P, Widdowson M, Kelley S (2010) Tectonic setting and timing of the final Deccan flood basalt eruptions. Geology 38(9):839–842. https://doi.org/10.1130/G31072.121)
22. Houston D, Werritty A., Bassett, D., Geddes, A., Hoolachan, A., & McMillan, M. (2011). Pluvial (rain-related) flooding in urban areas: the invisible hazard
23. https://www.deccanherald.com/india/india-lost-4-bn-to-extreme-weather-last-year-1241449.html
24. https://hpsdma.nic.in/
25. https://himachalpradesh.pscnotes.com/himachal-pradesh-polity/divisions-and-districts-of-himachal-pradesh/
26. Integrated Research and Action for Development https://www.irade.org/
27. https://hpsdma.nic.in/index1.aspx?lsid=72&lev=2&lid=14&langid=1
28. Grundfest E, Ripps A (2000) Flash floods, In: Floods. Parker DJ (Ed), 1, pp 377–390, Routledge, ISBN 978–0415172387, London and New York
29. UNDRR (2023), Annual Report 2022, United Nations Office for Disaster Risk Reduction (UNDRR)
30. UNDRR (2018) Annual report, Understanding hazard, vulnerability, and capacity assessments (HVCAs) in urban contexts: A 5-step process guide (consultation edition)
31. Thomas G, Vaze A (Insurance I. of I. (2018) Challenges in Insuring Flood-Prone Areas of Mumbai. 44)
32. Tiwari BA (2021) BlogURK Reimagining Cities in the Global South (www.urk.tiss.edu) A River Lost : The Concretising Landscape of Mumbai ' s Mithi. 1–8. 45)

33. Turner BL, Kasperson RE, Matsone PA, McCarthy JJ, Corell RW, Christensene L, Eckley N, Kasperson JX, Luers A, Martello ML, Polsky C, Pulsipher A, Schiller A (2003) A framework for vulnerability analysis in sustainability science. Proc Natl Acad Sci USA 100(14):8074–8079. https://doi.org/10.1073/pnas.1231335100
34. Zhu Y, Tian D, Yan F (2020) Effectiveness of entropy weight method in decision-making. Math Probl Eng 2020:1–5. https://doi.org/10.1155/2020/3564835
35. Sodhi, Balwinder, Tadinada, Prabhakar (2012) A simplified description of fuzzy TOPSIS

Issues, Options, and Strategies for Making Hill Habitat Inclusive, Safe, Resilient, and Sustainable

Jit Kumar Gupta

1 Introduction

Hilly areas play a crucial role in promoting rational growth and sustainable development, with significant implications for the environment, ecology, and culture of any region, state, or country. Despite occupying 18% of the nation's geographical area and housing a population of 70 million, hill areas face inherent dualities and contradictions. Rich in natural resources, they often struggle with poverty due to lower development levels. Human activities have increasingly damaged their ecology and environment, with unscientific exploitation and poorly planned projects threatening their sensitive ecosystems. The Stockholm Conference on Human Environment in 1972 highlighted the need for global efforts to mitigate environmental deterioration and promote rational and sustainable development in hill areas.

Major obstacles to the rational development of hill areas stem from post-independence developmental policies, particularly in agriculture and tourism, leading to extensive deforestation and ecological imbalances. Despite being labeled ecologically sensitive zones, effective strategies for their rational development are lacking. Recognizing the fragility of hill ecosystems, distinct policies, programs, and approaches are needed, integrating conservation measures into development plans.

Historically, British development efforts in the nineteenth century created towns, resorts, and infrastructure in hill areas, but post-independence urbanization and rapid tourism-driven expansion have brought congestion, pollution, and deforestation. Environmental problems in hill areas result from an unbalanced approach by policymakers and development agencies, neglecting the preservation of natural resources, biodiversity, and scenic beauty. It's imperative to plan and develop hill areas rationally, safeguarding resources, preserving natural beauty, and maintaining biodiversity through well-designed developmental strategies.

J. K. Gupta (✉)
IET Bhaddal, Chandigarh, India
e-mail: jit.kumar1944@gmail.com

V. Chettry et al. (eds.), *Geoinformatics for Hilly Regions*, Advances in 21st Century Human Settlements, https://doi.org/10.1007/978-981-95-0156-4_13

2 Issues, Challenges and Strategies for Hill Areas

From the snowy peaks of the Himalayas to the barren hills of the Aravallis, the unique characteristics and physical conditions of hill areas lead to diverse challenges and opportunities (Fig. 1). Common issues in these regions include deforestation, unsustainable resource exploitation, erosion, flooding, and agricultural inefficiencies. Addressing these challenges while promoting sustainable development aligns with the principles outlined by the World Commission on Environment and Development. State governments must establish robust planning mechanisms to ensure future growth adheres to sustainable practices, fostering economic prosperity and orderly development.

Understanding the complexities of development in hill states requires consideration of ecological preservation, resource management, community cohesion, and settlement structures. It's crucial to anticipate and mitigate potential threats to effectively address the challenges faced by hill areas. In the subsequent sections, we will delve into these matters in detail and explore strategies for tackling them.

2.1 Balancing Development and Environmental Preservation

Despite their ecological richness, hill areas continue to lag economically, with substandard living conditions despite abundant natural resources. This disparity

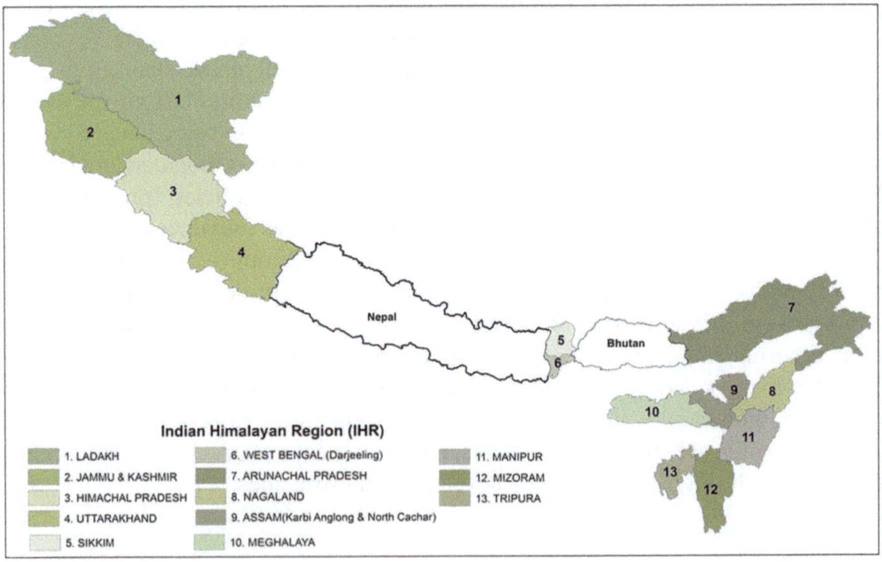

Fig. 1 Indian Himalayan region (IHR) [1]

stems from the haphazard and unscientific development approaches applied in these regions. To address this, comprehensive strategies must prioritize environmental conservation, sustainable land use, alternative energy sources, heritage preservation, planned tourism development, rational settlement systems, and resource utilization [2]. Development efforts should prioritize eco-restoration, eco-development, and eco-preservation, ensuring minimal disruption to existing resources, flora, fauna, and fragile ecosystems [3].

2.2 *Regional Planning for Sustainable Growth*

Diverse population distributions and density patterns in hill states pose significant planning and developmental challenges, necessitating varied options and strategies for rational and realistic progress. Embracing a regional approach proves valuable as Regional Planning offers the potential to balance conservation, preservation, and development in hill areas while safeguarding the ecology and environment. Utilizing existing physical, social, economic, and cultural indicators, hill states can be categorized into zones of varying sensitivity levels, emphasizing preservation in extremely/highly sensitive areas and focusing development in medium and low sensitivity zones. Ecologically degraded areas could be targeted for restoration and development based on their carrying capacity. Adopting this regional approach would facilitate the integrated development of urban and rural areas, optimizing resource utilization, resolving conflicts between authorities, reducing migration, and fostering sustainable, inclusive, safe, and resilient settlements.

2.3 *Rational Land Use for Hill Habitats*

In hill areas, challenges such as steep slopes, soil composition, biodiversity, vegetation, rock formations, and land structure pose significant obstacles to developing rational planning and development frameworks for state agencies. The limited availability of land for agricultural and non-agricultural purposes, particularly in eco-fragile zones, strains resources. Rampant deforestation to create land for development exacerbates the destruction of vital flora and fauna reservoirs. The pace of development often exceeds the sustainable capacity of these lands, with powerful interest groups exploiting vulnerable areas for profit. Effective land management, conservation, and finding a balance between agricultural and non-agricultural uses are essential for the sensible growth and development of hill states.

Strategic land use planning based on land structure, topography, capacity, and suitability is crucial for planned development and efficient utilization of land resources. Such planning could involve intensive afforestation on steep slopes, horticulture on moderate slopes, cultivation of food crops and human habitation in valleys and flatlands, and harnessing hydropower for electricity generation. Water resources could

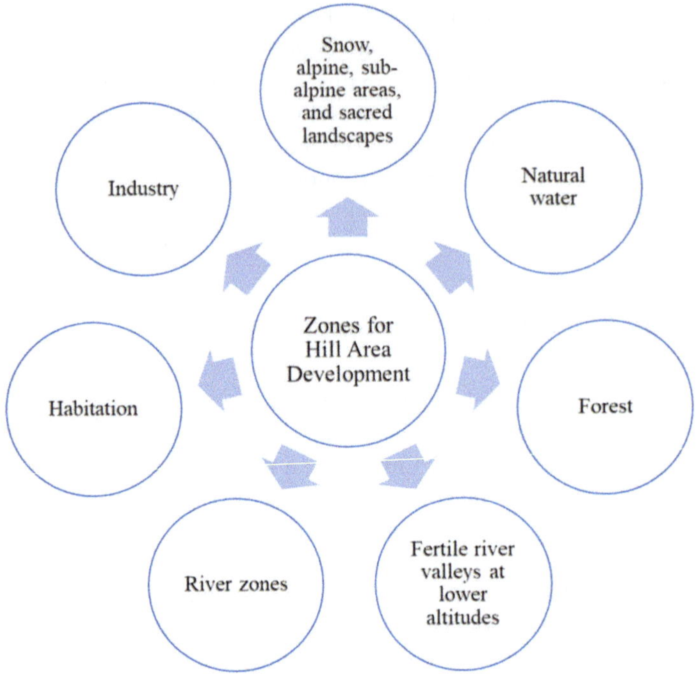

Fig. 2 Different zones for rationalising the hill area development recommended by the task force

also be utilized for irrigation, groundwater recharge, and enhancing vegetation cover in degraded areas.

To rationalize development in hill areas, the task force appointed by the Planning Commission of India proposes dividing hill states into distinct zones, each with specific activities [4] (Fig. 2).

- Snow, alpine, sub-alpine areas, and sacred landscapes would be designated as protected zones to maintain vital ecosystem services and preserve religious-cultural values.
- Zones with natural water sources (glaciers, rivers, lakes, and springs) would also be protected to safeguard water resources. Area's housing natural springs would become "Spring Sanctuaries" for preserving these resources.
- Forest zones would be conserved for environmental services and biodiversity, allowing for sustainable timber products, bamboo harvesting, and eco-tourism.
- Fertile river valleys at lower altitudes would be utilized for agricultural production, prohibiting conversion to other uses. Areas under shifting or terraced agriculture would promote unique crops, organic farming, horticulture, agroforestry, and improved agricultural practices.
- River zones suitable for decentralized electricity production would be designated to meet household and small industrial power needs.

- Zones designated for habitation, primarily urban spaces, would restrict construction on slopes above 30 degrees or in hazardous zones, springs, aquifer lines, and first-order streams. Efficient sewage and waste management systems would be mandatory.
- Industrial zones would be located in non-fragile areas, focusing on mountain-friendly activities using non-toxic, locally available materials, and prioritizing local employment and corporate social responsibility practices.

2.4 *Sustainable Industrial Growth*

The industry holds both positive and negative impacts on the economy, employment, development, and prosperity of any area. While it contributes to economic growth and job creation, it also consumes land and natural resources, generates waste, and impacts the environment and ecology, often leading to the creation of slums and shanty towns. Industrial growth in hill states must be carefully managed to avoid adverse effects on the environment and resources. Despite the abundant natural resources in hill areas, industrial development has been limited, with most industrial activities concentrated near the borders of developed states, causing spatial imbalances.

The prevailing pattern of industrial growth has harmed state resources, environment, and economy, with large-scale polluting industries being set up without adequate infrastructure. Industrial policies aimed at incentivizing growth have often led to the misuse of land and resources, further exacerbating environmental degradation. Non-polluting, hi-tech industries and agro-based industries using local produce should be encouraged in low eco-sensitive areas to promote employment and economic growth while minimizing environmental impact.

According to a task force appointed by the Planning Commission of India, industrial investments in hill states should be regulated to protect the region's fragility. Industrial development should be limited to selected areas, focusing on small and cottage industries linked to tourism and utilizing locally available raw materials. This approach can improve the local economy and provide employment opportunities for the local population.

2.5 *Fostering Eco-tourism for Hill Areas*

The relationship between hills and tourism is often positive, as people are drawn to explore and appreciate the unique climate, culture, ecology, and environment of hill areas. Tourism brings economic benefits, generates employment, promotes development, and alleviates poverty in tourist destinations. Therefore, leveraging tourism has become a priority for hill states to drive development, create wealth, and generate employment opportunities. However, tourism also has its drawbacks, impacting biodiversity, ecosystems, ecology, environment, culture, and development adversely.

Uncontrolled tourism has led to irreversible damage to valuable resources in hill areas, with poorly planned development and substandard infrastructure. Construction activities in ecologically fragile areas, disregard for carrying capacity, and violations of norms and building codes have further exacerbated environmental degradation. To address these issues, regulating uncontrolled tourism and related development is necessary to ensure harmony between tourism, environment, and development.

Tourism concentration in specific areas has also disrupted local culture, natural resources, ecology, and environment, leading to haphazard development and increased vehicular population in ecologically sensitive areas. Diversifying tourism destinations and promoting lesser-known areas can help distribute tourist footfalls, preserve vital resources, and stimulate economic growth and development in undeveloped regions.

Policies and programs for developing tourist infrastructure should be realistic, rational, objective, and aligned with the natural environment and surrounding resources. Local communities should actively participate in policy formulation to address conflicts between tourism and local livelihoods. Leveraging various types of tourism, such as heritage, religious, eco-tourism, adventure, cultural, rural, educational, wildlife, and wellness tourism, should be considered based on environmental sustainability, carrying capacity, and socio-cultural characteristics [5, 6].

Tourism planning should be integrated into overall local area planning and development strategies, guided by eco-tourism policies and national tourism policies. Emphasizing green tourism can promote sustainability by conserving resources, supporting local livelihoods, and reducing poverty while respecting cultural heritage and traditional values.

2.6 Disaster Preparedness and Mitigation Strategies

With earthquake-prone Zone-IV and Zone-V covering a significant portion of India's geographical area, the nation is highly susceptible to natural disasters triggered by seismic activity. Due to inherent vulnerabilities stemming from soil conditions, slope, topography, and fault lines, hill areas are particularly prone to disasters like earthquakes, landslides, and cloud bursts. States such as Himachal Pradesh, Jammu and Kashmir, Uttarakhand, Assam, Nagaland, and Meghalaya face heightened risks of such calamities [7]. Ensuring the safety of hill settlements against natural disasters is imperative, yet comprehensive vulnerability studies are often neglected prior to development projects in these regions. Consequently, when disasters strike, the lack of preparedness results in significant damage and loss of life, property, and infrastructure.

To enhance safety and resilience in hill areas, conducting detailed studies of physiography, topography, soil composition, slope stability, etc., is essential to identify stable and unstable zones. Only stable zones should be earmarked for physical development and human settlements, while unstable zones should be excluded to minimize vulnerability and disaster risk. Creating vulnerability atlases through comprehensive mapping of hill states' geographical areas is crucial for disaster preparedness and mitigation efforts.

As recommended by the task force established by the Government of India, implementing rational building regulations is critical for ensuring safety and disaster mitigation in hill areas. These regulations should govern permissible construction types, prioritize fire and earthquake safety measures, impose strict penalties for violations, mandate rainwater harvesting structures, establish norms for sewage disposal, waste management, and drinking water supply, designate refuge spaces in emergencies, install forecasting and early warning systems, and establish rapid response mechanisms. Such measures are essential for promoting safety and resilience, thereby mitigating the impact of disasters in hill areas.

2.7 Conservation and Preservation of Hill Heritage

Concepts of heritage areas, heritage zones and heritage cities/towns need adoption to protect the existing environment in heritage-rich areas. Adequate incentives need to be built to encourage and involve people and communities in preserving heritage, with stringent measures put in place to discourage damage and destruction of such areas. Development controls for heritage areas should be designed with sensitivity to maintaining and preserving the primary character of the area. Involving, educating and sensitising local people and communities about the need, role, relevance and importance of heritage, its preservation and conservation would remain critical and valuable for making heritage preservation/conservation a public-led mission rather than a government-promoted program.

Hill states, with their close connection to nature and unique cultural heritage, possess a wealth of valuable historical and natural treasures, including temples, palaces, bio-sphere reserves, lakes, forests, and wildlife sanctuaries. However, rapid population growth and unplanned development have led to the loss and extensive damage of this heritage. Unauthorized conversions, poor maintenance, uncontrolled tourism, and inadequate regulations have further contributed to the deterioration of these valuable assets. Encroachments and substandard development have marginalized and diluted the quality of heritage sites.

Recognizing the pivotal role of heritage in bridging the past and present, restoring community pride, promoting tourism, and driving economic development, conservation and preservation efforts become imperative. Any development strategy that overlooks heritage conservation risks long-term negative consequences. Hill states must prioritize heritage conservation through legal frameworks, plans, and development guidelines that address the root causes of damage to heritage sites.

Inspired by successful restoration projects like Machu Picchu in Peru and Bhaktapur Durbar Square in Nepal, hill states should adopt similar multidisciplinary approaches involving historical research, architectural conservation, and community engagement. Local expertise and indigenous knowledge should be leveraged to preserve cultural heritage while promoting sustainable development and local economic empowerment.

Establishing heritage areas, zones, and cities/towns can provide additional protection to these environments, with incentives in place to encourage community involvement in preservation efforts. Development controls should be sensitive to maintaining the area's character, and public education initiatives should raise awareness about the importance of heritage preservation.

Ultimately, heritage conservation should be a community-led initiative, with local people actively involved in preserving and safeguarding their cultural and natural heritage.

2.8 Enhancing Accessibility and Mobility in Hill Habitats

In hill states, effective transportation infrastructure remains a perpetual challenge due to the unique terrain and undulating topography. Roads, being the primary mode of access, are vital for mobility but are also vulnerable to uncontrolled development. Linear growth along roads, characterized by unplanned construction of shops, eateries, and other structures, leads to traffic congestion and limits future road widening efforts. To address this, regulations must be enforced to control development along roads, with strict penalties for violations. Revenue generated from permitted construction should contribute to road infrastructure upgrades.

The creation of efficient road, rail, and air transportation networks is essential for the rational development of hill states. A task force by the Planning Commission of India proposed loop railway lines for the Western Himalayan region and integration of road networks with rail and air services. Government initiatives, such as highway projects and airport development, aim to enhance connectivity and spur growth in hill states.

However, connectivity projects must be executed with environmental sensitivity. Mechanized transport for intercity access should be balanced with pedestrian-friendly and non-fuel-based mass transit systems for intra-city mobility. Environmental impact assessments are crucial to mitigate adverse effects on wildlife habitats and fragile ecosystems. For example, the construction of Nepal's East–West Highway required wildlife crossings and habitat restoration measures to minimize disruption to critical habitats and wildlife corridors. Prioritizing accessibility over mere mobility within cities can enhance the well-being of local populations while safeguarding the environment.

2.9 Building Design Strategies for Hill Habitats

Given the importance of energy for economic growth and development, coupled with concerns about energy consumption and environmental impact, addressing energy efficiency in construction is crucial. In hill areas, where energy-intensive methods are often necessary due to cold climates, innovative building design solutions are needed to reduce energy requirements.

Strategic site planning, such as constructing on Southern, Eastern, and Western slopes to maximize solar energy utilization, is essential. Passive design techniques like double-glazed windows, proper sealing, solar chimneys, and Trombe walls can help manage heat effectively and minimize energy loss. Drawing inspiration from traditional architectural techniques, such as rammed earth construction and passive solar design, can further promote energy efficiency in mountainous regions.

Using locally sourced materials and innovative building technologies is vital for sustainable construction. Building bye-laws should be revised to support energy-efficient designs, and retrofitting existing buildings is necessary to reduce energy consumption. Employing a combination of strategies and solutions, including optimizing space, incorporating greenery, and utilizing natural resources, can result in sustainable and energy-efficient buildings that are well-suited to hill environments.

3 Creating a Dedicated Cadre of Professionals

Looking at the physical, social, economic, cultural peculiarities, hill areas remain different and distinct from plain areas in any state and country; accordingly, hill areas would require specialised skills, training and expertise for planning human settlements; different architectural skills for designing the built environment and specialised engineering skills for the development of hill areas and making provision of the physical and social infrastructures. Although hills occupy more than one-sixth of the total area in the country, India does not have any specialised course in planning, designing and developing hill areas. Accordingly, all planning and designing of cities and buildings is carried out on proxy following the pattern used for planning the plain areas. It must be understood and appreciated that most ills plaguing the hill area have genesis in the non-availability of adequate and appropriately skilled manpower trained in the domain of hill area development [8]. Considering the larger context of rational and realistic development of hill areas, it will be critical that a pool of trained manpower, having expertise, understanding, knowledge and experience in the domain of hill areas planning, development and management, is created on priority.

Creating a pool of trained/skilled manpower, would require putting in place courses tailored to hill area planning and development and made operational. Urban local bodies must be empowered with skilled manpower to rationally and objectively address the issues the hill settlements face. Task force set up by the Planning

Commission of India to look into the issues and suggest options for hill area development has also recommended that; designing of mountain-specific courses and skilling planners, administrators, engineers, and social scientists, especially on all aspects of mountain ecology, geology, and marginal socio-economic conditions of the local population, should be considered as the priority area by the government.

4 Conclusion

Despite being a gift of nature to humanity, hill areas remain marginalised, neglected, muted and diluted in the parlance of planning, development, infrastructures, services, communication, economy, employment and productivity. In the process, large resources of hill areas are being misused, abused and destroyed in the name of development. Communication systems are being laid at the cost of local flora and fauna and damage to the very fabric/structure of the hill areas. Tourism is fast emerging as a major threat to the local environment, ecology, resources and culture. In the process, hill areas are losing their vitality, resources, valuable environment and ecology. The greed of speculation is leading to the mutilation of land resources, which is fast becoming a tradeable commodity. It must be appreciated that hill areas with a sensitive ecosystem require an equally sensitive, environment-friendly, resource-efficient, and scientific and state-of-the-art approach based on prevailing ground realities and local initiative involving local communities and resources/expertise, ensuring orderly growth and development of hill areas. For empowering hill areas and making optimum use of the potential of hill resources, it will be critical to leverage technology and use ICT for mapping available resources, understanding the impact of development, mapping physiography, topography, evolving short-term and long-term planning and development strategies; preparing master plans/development plans/local area plans; granting planning permissions for development; identifying vulnerable areas; mitigating and managing disasters; ensuring efficient delivery of services; promoting communication and ensuring overall growth and development of hill areas.

Leveraging Geographic Information Systems (GIS) and remote sensing technologies can significantly enhance the effectiveness of planning and development efforts in hill areas. For instance, the use of GIS-based spatial analysis tools allows planners to accurately assess terrain characteristics, identify environmentally sensitive areas, and develop informed land-use plans that minimize ecological disruptions. In Nepal, the use of drones equipped with high-resolution cameras has revolutionized terrain mapping and disaster management in mountainous regions. By capturing detailed aerial imagery, these drones enable authorities to rapidly assess damage following natural disasters, facilitating timely response and recovery efforts. Additionally, emerging technologies such as Building Information Modeling (BIM) offer advanced tools for sustainable building design and construction. BIM software allows architects and engineers to create digital models of structures, enabling accurate visualization, performance analysis, and energy-efficient design optimization. By embracing such innovative technologies, policymakers and practitioners can enhance the sustainability and resilience of development projects in hill habitats, ensuring their long-term viability and ecological integrity.

Task force report has recommended the need and urgency of reshaping national policies by embedding the "mountain perspective". Looking at the entire gamut of growth and development, hill areas must be valued, revered, preserved, protected and promoted for humanity, communities and the environment to promote global sustainability and achieve the agenda defined in the 17 SDGs.

References

1. Singh RK (2022) Data mining driven Himalayan knowledge network (HKN) for bridging science, policy and practice to foster sustainable development in the Indian Himalayan region. Int J Basic Appl Sci 11(4):135–140
2. Meshram DS, Chotani ML (2018) An approach to hill area development. In: Need for a comprehensive eco-strategy. V.K. Publisher, New Delhi, Uttrakhand
3. Sekar SP, Thirumeni M (2002) Planning strategies for hill stations in eco-sensitive zones. Spatio-Econ Dev Rec 9(2):35–37
4. India Water Portal Homepage (2011) Planning commission of India; report of the task force; problems of hill states and hill areas and ways to ensure that they do not suffer in any way because of their peculiarities. https://www.indiawaterportal.org/articles/problems-hill-states-and-hill-areas-and-ways-ensure-they-do-not-suffer-any-way-because. Accessed 16 Oct 2023
5. Eco-Tourism Policy, Ministry of Tourism, Government of India (1997)
6. Draft National Tourism Policy, Ministry of Tourism, Government of India (2022)
7. Gupta JK. Planning for sustainable development of hill state of Himachal Pradesh
8. Lotusarise Homepage. Hill area development. https://lotusarise.com/hill-area-development-programme-upsc. Accessed 5 Mar 2021

Academic Contribution and Policy Perspectives: Geoinformatics for Sustainable Human Settlements in Hilly Regions

Bharat Dahiya, Vishal Chettry, Harsimran Kaur, and Sarbeswar Praharaj

1 Academic Contribution and Policy Perspectives

This concluding chapter provides an overview of the academic contribution made by the chapters included in this anthology. In doing so, it also highlights the new policy perspectives recommended therein.

The introductory chapter of the book by Vishal Chettry, et al. [1] provided a brief background to the geoinformatics for sustainable human settlements in hilly regions and underscored the importance of developing sustainable human settlements in hilly regions. After outlining geoinformatics and its components, it highlighted the significance of the application of geoinformatics to human settlements and sustainable development. Then, it provided a brief introduction to the chapters included in the book. In its concluding remarks, it highlighted the importance of applying geoinformatics towards improving knowledge of hilly regions and offering distinct approaches for their management and sustainable development.

The chapter titled 'Sustainable Urban Development in Hilly Regions: Concepts, Evolution, and Challenges' by Harsimran Kaur, and Akhil Nawani [2] offers significant academic and policy contributions. Firstly, it contributes to academic discourse

B. Dahiya (✉)
School of Global Studies, Research Center for Sustainable Development and Innovation, Thammasat University, Bangkok, Thailand
e-mail: bharat@sgs.tu.ac.th

School of Public Leadership, Stellenbosch University, Stellenbosch, Western Cape, South Africa

V. Chettry · H. Kaur
Department of Architecture, Planning and Design, IIT(BHU), Varanasi, India

S. Praharaj
Department of Geography and Sustainable Development and School of Architecture, University of Miami, Coral Gables, FL, USA

V. Chettry et al. (eds.), *Geoinformatics for Hilly Regions*, Advances in 21st Century Human Settlements, https://doi.org/10.1007/978-981-95-0156-4_14

by providing a comprehensive understanding of the role of geoinformatics in addressing the unique challenges faced by hilly terrains. By explaining the capabilities of remote sensing in acquiring spatial data and geographic information systems (GIS) in integrating and analysing this data, the chapter enhances knowledge in the field of urban planning and resource management. The exploration of strategies for sustainable development, such as identifying high-risk areas, land use planning, watershed management, and infrastructure planning, adds depth to academic literature by offering practical solutions grounded in geoinformatics technology. Moreover, the inclusion of case studies and examples demonstrates the real-world applicability of these strategies, enriching academic discussions with empirical evidence. From a policy perspective, the chapter provides valuable insights for policymakers and urban planners in hilly regions. By highlighting effective approaches and best practices enabled by GIS and remote sensing, the chapter informs policy formulation and implementation aimed at promoting sustainable urban development. Policymakers can draw upon the recommendations and considerations outlined in the chapter in order to design evidence-based strategies tailored to the specific geographical and environmental characteristics of hilly areas. Overall, the chapter's academic and policy contributions advance knowledge and provide practical guidance for sustainable urban development in hilly regions, fostering resilience, environmental conservation, and social inclusivity.

In their chapter, 'Land Use Planning in Hill Cities of India and Abroad: A Comparative Study', Swasti Sharma, Ashutosh Saini, Bhavna Shrivastava, Ashwani Kumar [3] conduct an in-depth study of hill areas with a special focus on best land use practices in the hill towns at both national and international level. The land parcels in the hills are scarce and the development is happening at a rapid pace. The hilly regions face various issues and challenges due to inappropriate land uses and lack of proper land use planning. The study of three planned Indian hill towns, Bilaspur, Lavasa and New Tehri, aimed to look into the planning concepts used in these towns but the procedure or the method of land allocation process was not mentioned in these documents. Land use planning in the Indian hills is ineffective due to the paucity of information, lack of data, improper documentation, the absence of a scientific basis for the allocation of land uses, and the lack of a standard method for allocating land uses. In comparison, the hill cities of Thimphu (Bhutan), Kathmandu (Nepal), and Santa Clarita, Anchorage, and Boone County (United States) demonstrate that systematic procedures are being followed along with proper documentation practices to ensure planned development. Therefore, with reference to the mentioned case examples, the study contributes to understanding the need for systematic land use planning practices in the hilly regions of India. It further emphasises the important need to work out a logical framework for appropriate moulding of this process in the overall societal interest of having contextual sustainable development in these eco-sensitive areas.

The chapter titled, 'Sustainable land-use planning in environmental sensitive Hill city Gangtok, India', Sarika Bahadure, et al [4] have made an effort to devise a methodology for land suitability analysis by amalgamating physical and social parameters specific to a hill city. The study has both academic and policy contributions as follows. First, the guided study was undertaken by urban planning students. Such studies help to sensitize future urban planners for a sustainable future and make them aware of the intricacies of hill cities. It makes them understand the dynamics of city planning and the manifold layers that influence an urban planning decision. It guides the students those spatial decisions have influences on not only physical but also social aspects in an urban area. Second, the study contributes to a robust framework for conducting land suitability analysis for a hill city through the identification of pertinent parameters and methodology for data collection. The weights of each indicator provide an insight of the relative importance of each indicator in a hill city. Also, the ranking system of each indicator (ranging from 0 to 3) offers guidance of the performance of the said indicator in a city. Hence, the scope of improvement can be traced with the help of such an evaluation system. Third, this is a GIS-based study and, hence, it provides flexibility in incorporating future data for ongoing progression studies or retrospective analyses to track the evolution of land suitability over time. Such methods thus always help in a continuous evaluation of city land use suitability. Hilly areas are highly vulnerable to disaster risks and hazards; hence such flexible methods are appropriate for contextual analysis. Finally, the study can be linked directly to the spatial planning of the city and helps to develop proposals for sustainable urban planning. The study therefore provides short-term, medium-term, and long-term proposals for the hill city of Gangtok based on the analysis. Accordingly, this study serves as a blueprint for other hill cities, guiding them towards informed decision-making processes for sustainable urban planning endeavours.

In their chapter, 'The impact of culture, climate, topography, and geology on settlements in hilly areas: A case of Kandad, India', Sameer Gujar et al. [5] conduct sustainable settlement analysis of Kandad Village, located in Chakrata sub-district, a part of Dehradun district in Uttarakhand state, which has significant academic and policy implications. By employing GIS to scrutinize various facets of the village's landscape and settlement patterns, it offers valuable insights into the challenges and opportunities associated with sustainable development in mountainous regions. Through geographical and topological analyses, the chapter identifies challenges like natural hazards, providing crucial information for disaster risk reduction strategies and land-use planning policies. The examination of built-up change over time sheds light on evolving land-use dynamics, informing decision-makers about the impacts of development interventions and the need for adaptive planning measures. Further, the documentation and analysis of traditional construction techniques contribute to preserving traditional knowledge systems and promoting culturally appropriate and resilient building practices. This aspect of the study not only underscores the importance of traditional knowledge and wisdom in sustainable development, but also offers practical lessons for policymakers and practitioners working in similar contexts globally. Furthermore, by integrating GIS-based findings with traditional construction documentation, the chapter emphasises the interdisciplinary nature of sustainable settlement planning and highlights the significance of holistic approaches that incorporate both scientific and indigenous knowledge systems. Overall, the chapter serves as a valuable resource for academia, policymakers, and practitioners involved in sustainable development efforts, offering actionable insights and methodologies to foster resilience and sustainability in mountainous regions worldwide.

In the next chapter, Parshant Rehal [6] examine the accessibility for senior citizens in human settlements in hilly regions and provide insights on the related challenges and opportunities for sustainable mobility. As the older adult population numbers are expected to rise in the coming decades due to population ageing in many countries, including India, cities and urban environments in developing countries will need to enhance their capabilities for age-care, health facilities, and socio-economic support systems. Although the benefits of an active lifestyle on well-being are evident in ageing studies, similar initiatives towards building age-friendly settlements, especially in the context of hill regions, are in their early stages. This chapter, therefore, attempts to expand the existing field of knowledge by focusing on the relationship between older adults' daily outdoor activities and the urban built environment in neighbourhoods, towns, and cities in hilly regions. Older adults residing in urban hill settlements often experience various challenges to active mobility due to the complex topographic conditions, which can lead to physical and psychological barriers among them toward independent and successful ageing experiences. The chapter thus helps identify the barriers towards outdoor mobility among older adults caused by inadequate design standards and planning regulations or topographical constraints. The chapter calls for implementing strategies towards enhancing accessibility and mobility linkages between the home and outdoor destination of daily activities as an urban planning and design initiative for improving overall life satisfaction among older adults. The chapter provides knowledge inputs that could be helpful to policymakers, planning and design experts, and academicians towards framing strategies for building sustainable and age-friendly hill settlements with the goal of enhancing social connectedness, functional independence, and quality of life among older adults.

The chapter titled 'Assessment of urban growth dynamics in Srinagar, Jammu and Kashmir, India' by Sneha, Vishal Chettry et al. [7] offers significant academic and policy contributions in understanding the urban growth dynamics and governance challenges in Srinagar, and thereby informs future research directions and policy interventions. Academically, the utilization of geospatial techniques to analyse urban growth patterns provides valuable insights into the spatial dynamics of development. By tracing the evolution of settlement patterns over nearly a century, the chapter contributes to a deeper understanding of the historical trajectory of urbanisation in Srinagar. This analysis sheds light on the factors driving urban expansion and identifies areas of underdevelopment, laying the groundwork for further research into the socio-economic and environmental implications of urban growth. The examination of governance frameworks and policy interventions offers critical insights into the effectiveness of urban planning strategies in Srinagar. The chapter highlights the challenges pertaining to political instability and regulatory enforcement and underlines the need for more robust governance mechanisms to guide sustainable development. By critically assessing past master plans and governance structures, this research informs future policy decisions aimed at fostering inclusive and balanced urban growth. With regard to policy, the findings of this chapter provide inputs for urban planning and development initiatives in Srinagar and beyond. By identifying areas of underdevelopment and spatial inequalities, policymakers can prioritize resource allocation and infrastructure investments to promote equitable growth. Moreover, the emphasis on effective governance mechanisms aligns with broader policy objectives of enhancing administrative efficiency and responsiveness to urban challenges. The chapter's discussion on the emergence of 'Srinagar Smart City' initiative signifies a paradigm shift towards sustainable urban development. By integrating traditional art, culture, and sustainability principles into urban planning, the Srinagar Smart City initiative represents a pioneering effort to address the multifaceted challenges of urbanisation in Srinagar. This highlights the potential for innovative policy interventions to foster socio-economic inclusivity and enhance residents' quality of life, ultimately contributing to the broader objectives of sustainable urban development in the region.

Nivedita [8] chapter, 'Satellite Remote Sensing Techniques for Air Quality Monitoring in the Mid-Brahmaputra Valley of Northeast India', significantly contributes to the understanding of aerosol dynamics in a climatically and topographically distinct region. The research addresses a critical gap in long-term atmospheric particle monitoring in northeast India—a region often overlooked in mainstream air quality assessments. By integrating ground-based measurements using Beta Attenuation Monitors and Aethalometers with satellite-derived aerosol optical depth (AOD) data from MODIS (Terra), the study successfully captures the diurnal and seasonal variations of particulate matter and black carbon over a rural hillock-dominated landscape. This approach offers a cost-effective, scalable alternative to conventional ground monitoring, especially in regions with limited infrastructure. Importantly, the findings highlight the role of meteorological variability in influencing particle behaviour—essential for interpreting aerosol-climate interactions in the Brahmaputra Valley. The research methodology utilised in the chapter can inform the development of regional aerosol models, air quality forecasting systems, and evidence-based climate action plans. From a policy perspective, this research emphasises the necessity of incorporating satellite remote sensing tools into national and regional air quality management strategies. It supports the deployment of hybrid monitoring networks combining satellite and in-situ data to bridge spatial gaps in pollution surveillance. Moreover, the dataset generated can serve as a scientific foundation for policies aimed at reducing exposure to particulate pollution, particularly in ecologically sensitive and underserved areas of Northeast India. The findings also hold potential for integrating air quality indicators into health and environmental risk assessments, reinforcing public health resilience and climate adaptation frameworks.

The chapter 'Spatio-Temporal Comparative Study of Aerosol Optical Depth, Aerosol Index, and Planetary Boundary Layer Height During and After COVID-19 Lockdown over North Eastern Region of India', by Divyanshu Saini et al. [9], contributes to both academic understanding and policy formulation in the realm of air quality management, particularly in the context of India's response to the COVID-19 pandemic. By analysing Aerosol Optical Depth (AOD), Aerosol Index (AI), and Planetary Boundary Layer Height (PBLH) using data from MODIS Terra and OMI Aura, the study provides novel insights into the complex dynamics of aerosol pollution and atmospheric conditions. Academically, the chapter advances knowledge by revealing temporal and spatial patterns in aerosol levels during and after the COVID-19 lockdown periods. Notably, the identification of a concurrent increase in AOD and PBLH challenges conventional understanding, underscoring the need for a more nuanced approach to studying aerosol composition and its impact on air quality. Moreover, the study sheds light on the specific regional dynamics within India, particularly in the Northeast region, where positive AI values suggest unique challenges in aerosol management. From a policy perspective, the chapter offers actionable insights for policymakers and stakeholders. By demonstrating the impact of human activity on aerosol concentrations and air quality, the research underscores the importance of implementing effective clean air action plans. The findings emphasise the need for comprehensive strategies that account for seasonal variations, regional disparities, and the interplay between anthropogenic factors and atmospheric conditions. Such nuanced understanding is crucial for formulating evidence-based policies aimed at mitigating air pollution, protecting public health, and achieving sustainable development goals. Overall, this chapter bridges the gap between academic research and policy implementation, providing valuable guidance for addressing the complex challenges of air quality management in the wake of the COVID-19 pandemic.

In their chapter 'Geoinformatics for Climate Change Vulnerability Assessment in the Western Himalayan Cities', Ranjeet Verma et al. [10] focus on the integration of geoinformatics tools and techniques in assessing climate change vulnerability and provide insights into the localized impact of climate change in Western Himalayan cities. Climate change severely impacts these cities, which face numerous challenges, including deforestation, rapid population growth, landslides, flash floods, and water scarcity. These challenges may be exacerbated in the coming decades due to climate change. Geoinformatics and its components could significantly contribute to understanding the vulnerability of these cities to climate change and thus aid in mitigating the associated risks. The chapter emphasizes the policy relevance of geoinformatics-based vulnerability assessments in formulating climate change adaptation and mitigation strategies by identifying vulnerable areas. By leveraging this information, policymakers and urban planners could prioritise resources to develop resilience frameworks and reduce the impacts of climate change. This approach could guide decision-makers toward effective, sustainable, and inclusive strategies for mitigating climate change-induced vulnerability in Western Himalayan cities. Moreover, this chapter draws multidisciplinary insights from environmental science, geography, urban planning, and sociology to develop a deeper understanding of the complexity of climate change vulnerability. This multidisciplinary approach enriches the vulnerability assessment, providing researchers and academicians with a holistic understanding of the challenges and opportunities for adaptation and mitigation strategies in Western Himalayan cities. Furthermore, the chapter contributes to the academic literature on climate change adaptation by showcasing the application of geoinformatics and its components in vulnerability assessment, particularly in mountainous cities. It also offers evidence-based methodological insights for researchers and academicians to utilize in future research endeavours in the field of climate change.

The chapter 'Assessing and Mapping Tourism Vulnerability using Remote Sensing and Spatial Analysis: A Case of Shimla District in Himachal Pradesh, India', by Keerti Manisha et al. [11], makes a significant academic and policy contribution by addressing the gap in empirical studies pertaining to the assessment of tourism vulnerability in Shimla district, Himachal Pradesh. This study offers a novel approach to evaluating the intricate dynamics of tourism vulnerability in a geographically diverse and ecologically sensitive region by utilizing remote sensing and spatial analysis techniques coupled with the 'Analytic Hierarchy Process' as a multi-criteria decision mapping method. By systematically mapping and analysing 18 indicators classified under the 'Exposure, Sensitivity, and Adaptive Capacity' criteria, the chapter provides a comprehensive understanding of the factors influencing tourism vulnerability in Shimla district. The thematic maps and composite vulnerability map generated through this study enhance the scholarly discourse on tourism vulnerability assessment and offer practical insights for policymakers and stakeholders. The identification of vulnerable areas, along with their associated risk factors, enables informed decision-making in tourism planning and development. Moreover, the policy implications derived from this research, such as enhancing adaptive capacity through capacity building and stakeholder engagement, hold significant relevance for fostering sustainable tourism practices and mitigating the negative impacts of tourism on the natural and cultural heritage of Shimla district. This chapter thus serves as a valuable resource for academia, policymakers, and practitioners seeking to promote resilient and sustainable tourism development in hill areas like Shimla.

In their chapter, 'Safeguarding the Western Himalayas: Sustainable Disaster Management in the Hilly Regions of Himachal Pradesh', Shubham Sharma et al. [12], make notable academic and policy contribution to the discourse on disaster management, particularly within the unique context of the geographical and climatic challenges in Himachal Pradesh. Through the implementation of a rigorous methodology and a comprehensive review of existing literature, valuable insights are offered into the multifaceted nature of disaster risk and vulnerability in the region. The analysis deepens the understanding of the specific hazards faced by Himachal Pradesh while providing methodological approaches that can be adapted and applied to similar regions facing same challenges worldwide. The authors significantly contribute to disaster management research by employing GIS technology alongside the 'Technique for Order of Preference by Similarity to Ideal Solution' (TOPSIS) methodology. Through the integration of GIS, nuanced insights into the physical geography and landscape features of Himachal Pradesh are gained, enhancing the accuracy and precision of the analysis. GIS facilitates the mapping and visualization of vulnerable areas, furnishing decision-makers with spatially informed understanding crucial for targeted interventions and resource allocation. Moreover, utilizing the TOPSIS methodology enhances the analysis by providing a thorough evaluation of physical vulnerability based on multiple criteria such as elevation, land use, and drainage systems. The systematic ranking of areas according to their susceptibility to flooding provided by TOPSIS offers valuable insights for informed decision-making in disaster risk management strategies. This integrated approach not only advances the understanding of flood vulnerability in the Kullu district, but also demonstrates the scalability and potential applicability of the methodology to broader geographical contexts within Himachal Pradesh. The adept utilization of GIS and TOPSIS highlights the utility of employing innovative techniques for advancing the field of disaster management, which could contribute to more effective and sustainable disaster mitigation and adaptation strategies in the region.

In the chapter titled, 'Issues, Options and Strategies for making Hill Habitat Inclusive, Safe, Resilient and Sustainable', Jit Kumar [13] sheds light on the overlooked yet crucial aspects of planning and development in hill areas. By delineating the unique challenges faced by hill habitats and proposing tailored strategies for addressing them, the chapter fills a gap in existing literature and policy discourse. It emphasizes the need for specialized training and expertise in hill area development, advocating for the creation of a dedicated cadre of professionals. The incorporation of real-life examples in the chapter adds practical applicability to its recommendations, which could facilitate knowledge transfer and capacity building. From a policy perspective, the chapter calls attention to the imperative of integrating environmental impact assessments, community engagement strategies, and climate resilience measures into hill area development policies. By advocating for inclusive and sustainable approaches grounded in ethical considerations, the chapter provides actionable insights for policymakers and practitioners alike, with the aim to foster balanced and resilient development in hill areas.

2 Conclusion

Through the above reflection, we found that a common thematic focus of the studies contained in the various chapters was the application of geoinformatics in analysing multifarious challenges faced by and developing a better understanding of the human settlements related problems in the hilly regions. These chapters examined a multiplicity of problems, such as land use planning, urban growth dynamics, air-quality monitoring, accessibility for senior citizens, climate change vulnerability assessment, assessing and mapping tourism vulnerability, and sustainable disaster management. The studies contained in this anthology have made important academic contribution and provided fresh policy perspectives for improving human settlements in hilly regions. All of this has provided a renewed emphasis on the significance of geoinformatics and its application in addressing different problems faced in the process of working towards sustainable human settlement development in hilly regions.

References

1. Chettry V, Kaur H, Praharaj S, Dahiya B (2025) Introduction: geoinformatics for sustainable human settlements in hilly regions. In Chettry V, Kaur H, Praharaj S, Dahiya B (Eds., Geoinformatics Hilly RegNs: Sustain Hum Settl. Springer

2. Kaur H, Nawani A (2025) Sustainable urban development in hilly regions: concepts, evolution, and challenges. In Chettry V, Kaur H, Praharaj S, Dahiya B (Eds), Geoinformatics Hilly Reg.: Sustain. Hum. Settl. Springer
3. Sharma S, Saini A, Shrivastava B, Kumar A (2025) Land use planning in hill cities of india and abroad: a comparative study. In Chettry V, Kaur H, Praharaj S, Dahiya B (Eds), Geoinformatics Hilly Reg.: Sustain. Hum. Settl. Springer
4. Bahadure S, Kamble TM, Suthar ML (2025) Sustainable land-use planning in environmental sensitive Hill city Gangtok, India. In Chettry V, Kaur H, Praharaj S, Dahiya B (Eds), Geoinformatics Hilly Reg.: Sustain. Hum. Settl. Springer
5. Gujar S, Kulkarni V, Maurya A, Bahale S (2025) The impact of culture, climate, topography, and geology on settlements in hilly areas: A case of Kandad, India. In Chettry V, Kaur H, Praharaj S, Dahiya B (Eds), Geoinformatics Hilly Reg.: Sustain. Hum. Settl. Springer
6. Rehal P, Chani PS, Atreya S, Nawani A, Kaur H (2025) Accessibility for senior citizens in hill settlements: challenges and opportunities for sustainable mobility. In Chettry V, Kaur H, Praharaj S, Dahiya B (Eds), Geoinformatics Hilly Reg.: Sustain. Hum. Settl. Springer
7. Sneha, Chettry V, Yadav A (2025) Assessment of urban growth dynamics in Srinagar (J&K), India. In Chettry V, Kaur H, Praharaj S, Dahiya B (Eds), Geoinformatics Hilly Reg.: Sustain. Hum. Settl. Springer
8. Barman N (2025) Satellite remote sensing techniques for air quality monitoring in the Mid-brahmaputra valley of northeast India. In Chettry V, Kaur H, Praharaj S, Dahiya B (Eds), Geoinformatics Hilly Reg.: Sustain. Hum. Settl. Springer
9. Saini D, Lataye DH (2025) Spatio-Temporal comparative study of aerosol optical depth, aerosol index, and planetary boundary layer height during and after COVID-19 lockdown over North Eastern Region of India. In Chettry V, Kaur H, Praharaj S, Dahiya B (Eds), Geoinformatics Hilly Reg.: Sustain. Hum. Settl. Springer
10. Verma R, Kaur H, Dwivedi A (2025) Geoinformatics for climate change vulnerability assessment in the western himalayan cities. In Chettry V, Kaur H, Praharaj S, Dahiya B (Eds), Geoinformatics Hilly Reg.: Sustain. Hum. Settl. Springer
11. Manisha K, Singh I, Jain M (2025) Assessing and mapping tourism vulnerability using remote sensing and spatial analysis techniques: a case of shimla district in Himachal Pradesh, India. In Chettry V, Kaur H, Praharaj S, Dahiya B (Eds), Geoinformatics Hilly Reg.: Sustain. Hum. Settl. Springer
12. Sharma S, Sharma S, Sharma S (2025) Safeguarding the western himalayas: sustainable disaster management in the hilly regions of Himachal Pradesh. In Chettry V, Kaur H, Praharaj S, Dahiya B (Eds), Geoinformatics Hilly Reg.: Sustain. Hum. Settl. Springer
13. Gupta JK (2025) Issues, options and strategies for making hill habitat inclusive, safe, resilient and sustainable. In Chettry V, Kaur H, Praharaj S, Dahiya B (Eds), Geoinformatics Hilly Reg.: Sustain. Hum. Settl. Springer

MIX
Papier aus verantwortungsvollen Quellen
Paper from responsible sources
FSC® C105338

If you have any concerns about our products,
you can contact us on
ProductSafety@springernature.com

In case Publisher is established outside the EU,
the EU authorized representative is:
Springer Nature Customer Service Center GmbH
Europaplatz 3, 69115 Heidelberg, Germany

Printed by Libri Plureos GmbH
in Hamburg, Germany